T0191775

Birds as Useful Indicators of High Nature Value Farmlands

Federico Morelli • Piotr Tryjanowski
Editors

Birds as Useful Indicators of High Nature Value Farmlands

Using Species Distribution Models as a Tool
for Monitoring the Health of Agro-ecosystems

Editors
Federico Morelli
Faculty of Environmental Sciences,
    Department of Applied
    Geoinformatics and Spatial
    Planning
Czech University of Life Sciences
    Prague
Kamýcká 129, Czech Republic

Piotr Tryjanowski
Institute of Zoology
Poznań University of Life Sciences
Poznań, Poland

ISBN 978-3-319-84366-7   ISBN 978-3-319-50284-7   (eBook)
DOI 10.1007/978-3-319-50284-7

Acknowledgments

We thank A. Wuczyński, C. Korkosz, F. Pruscini for pictures and T. Hartel for the graph and fruitful discussion on many aspects presented in the book. We would also thank our friends for help with data collection in the field: Z. Kwieciński, L. Jankowiak, M. Tobolka, M. Antczak and M. Hromada (Poland); F. Pruscini, C. Rubolini, A. Suzzi Valli and F. Martinelli (Italy).

Contents

1 Introduction   1
   Federico Morelli, Yanina Benedetti, and Piotr Tryjanowski

2 Spread of the Concept of HNV Farmland in Europe:
   A Systematic Review   27
   Yanina Benedetti

3 Identifying HNV Areas Using Geographic Information
   Systems and Landscape Metrics   37
   Petra Šímová

4 Suitable Methods for Monitoring HNV Farmland Using
   Bird Species   53
   Piotr Tryjanowski and Federico Morelli

Part I   Case Studies

5 Case Study 1. Bird as Indicators of HNV: Case Study
   in Farmlands from Central Italy   71
   Federico Morelli, Leszek Jerzak, and Piotr Tryjanowski

6 Case Study 2. Birds as Indicators of HNV: Case Study
   in Portuguese Cork Oak *Montados*   89
   João E. Rabaça, Luísa Catarino, Pedro Pereira, António Luís,
   and Carlos Godinho

7 Case Study 3. Using Indicator Species Analysis IndVal to Identify
   Bird Indicators of HNV in Farmlands from Western Poland 107
   Piotr Tryjanowski and Federico Morelli

8 Discussion and Final Considerations 115
   Federico Morelli and Piotr Tryjanowski

About the Editors

Federico Morelli is a Quantitative Ecologist, currently working as Researcher at the Czech University of Life Sciences (Prague, Czech Republic). He has been involved in several European projects modelling the impact of land use and climate change on spatial distribution of biodiversity. The main focus of his research interests is related to the response of species distribution models to multiscale predictors, the effects of landscape metrics on biodiversity patterns, the development and test of several bioindicators and the general topics of macroecology.

Piotr Tryjanowski is a professor and director of the Institute of Zoology at Poznan University of Life Sciences. His research interest includes mainly ecology of vertebrates, with particular attention to the effect of climate and environmental changes on birds and mammals. For instance, since more than 25 years he studies farmland bird populations in Poland.

Chapter 1
Introduction

Federico Morelli, Yanina Benedetti, and Piotr Tryjanowski

Abstract The decline of biodiversity in the agro-ecosystems: causes and consequences. The main differences between Western and Eastern Europe agriculture – the role of history. The development of conservation tools in Europe: from the network of protected areas and Nature2000 approach to the High Nature Value areas identification. Theories for fragmented landscapes and the application of ecological terminology for agricultural landscapes in Europe. Approaches useful to identify and characterize HNV farming systems. The HNV as support for biodiversity and public goods.

Keywords Farmland bird decline • Biodiversity conservation • Farming intensification • Protected areas • Natura2000 network • High nature value farming • HNV

1.1 The Decline of Biodiversity in the Agro-ecosystems

The agriculture is a dominant form of land use on the world's terrestrial surface, accounting for more than 40 % of land use coverage [1, 2]. In Europe, agricultural landscapes are artificial mosaics of different land use types, and represent one of the most common habitat. But during the last few decades the agricultural landscapes have been subject to a rapid and large-scale change, caused mainly by the intensification and mechanization of agricultural activities [3–8]. The agricultural intensification is devoted to feed the growing world population. However, this process is

F. Morelli (✉)
Faculty of Environmental Sciences, Department of Applied Geoinformatics and Spatial Planning, Czech University of Life Sciences Prague, Kamýcká 129, CZ-165 00 Prague 6, Czech Republic
e-mail: fmorellius@gmail.com

Y. Benedetti
Centro Naturalistico Sammarinese, via Valdes De Carli 21, 47893 Borgo Maggiore, San Marino

P. Tryjanowski
Institute of Zoology, Poznań University of Life Sciences, Wojska Polskiego 71 C, PL-60-625 Poznań, Poland

considered one of the chief drivers of worldwide biodiversity decline [4, 9, 10]. Agricultural intensification occurs mainly at two different spatial scales: local scale -e.g.
increased use of agrochemicals or pesticides [6] and landscape scale -e.g. destruction of semi-natural and marginal habitats [11]. The latter affects the marginal and
unproductive elements of farmland removing shrubs, hedgerows, isolated trees and
uncultivated patches. The marginal and unproductive elements of farmland landscapes, such as shrubs, hedgerows, isolated trees and uncultivated patches, constitute key habitats for many species, for nesting, feeding and protection [12–14], as
well as providing ecological corridors [15, 16] and to increase and maintain the
plant communities diversity [17].

Scientists currently recognize how biodiversity loss can negatively impact on
humanity in many different ways [18, 19]. The threat of biodiversity is a threat for
humans for the simple reason that biodiversity is the main driver of ecosystem
functioning where humans are living [20]. Then, the conservation of the biodiversity in the countryside is essential not only for intrinsic reasons, but also for very
pragmatic reasons related to human benefits [21]. Mankind benefits from a range of
services provided by nature [22]. From the many benefits that agriculture may
obtain, some of them are directly provided by wild organisms such as pollinators or
pest control in croplands provided by natural enemies [23–25].

However, the loss of biodiversity in agricultural landscapes is not uniform across
Europe, and need, then, more deep understandings. Because the diversity of rural
landscapes across the continent plus the inadequacy of existent datasets on biodiversity, can represent an impediment to the develop of a common strategy for all
European countries [2].

1.2   Differences Between Western and Eastern Europe
Agriculture – The Role of History

HNV areas are not equally distributed over Europe, but even intensity of studies on
HNV farmland on birds strongly differs between areas, both in local as well as
broad continental scale. Differences are connected mainly with low-intensity agricultural land use in Europe with has created many unique and species-rich assemblages [26, 27]. Paradoxically a large proportion of European species are now
dependent over much of their ranges on this form of human disturbance [28]. However, the industrialization of agriculture has, directly and indirectly, caused a
dramatic impoverishment of the avifauna and flora compared to the situation a
century ago [29]. This has contributed not only to the current biodiversity crisis in
Europe as a whole, but also to the decline in ecosystem services such as crop
pollination and biological pest control [29]. Among suggested solutions to save
farmland bird biodiversity are especially dedicated conservation schemes, with idea
to mitigate the impacts of intensive farming, and to the support of low-intensity
practices on existing high nature value (HNV) farmland. HNV farmland is present

throughout Europe, although it is often restricted to upland or other areas difficult to farm, particularly in Northern and Western Europe (EEA, 2004). Eastern and Southern Europe, in contrast, generally have lower average levels of land use intensity, and healthy populations of many species declining or endangered in the north-west persist here [26, 27]. Similarly, the farmland environment in Central-Eastern Europe is generally still more extensive than in Western Europe and a larger proportion of people still live in rural areas [26] what in consequences generating different conditions for both, organisms living in agricultural areas, as well as human societies and economy. Therefore to protect declining populations living in farmland, detailed knowledge on both species and communities level is necessary [26]. It seems nearly obvious since last few years [26, 30], however, due to scientific tradition and availability of funding, the definitely majority of studies have been carried out in Western Europe [27]. In consequence this provokes a question: are findings obtained in western conditions useful to identify the fate of farmland bird biodiversity in Central-Eastern Europe? Or even more general: how results obtained in particular local conditions can be generalized for understanding farmland functioning over the whole continent? Tryjanowski et al. [26] in a review argued that is not necessary. On the other hand is easy to say that we need more detailed studies, but the intention is also to show potential benefits (even economical) from develop of this kind of study. Tryjanowski et al. [26] provided statistical evidences that agriculture differs between western (WE) and central-eastern (CEE) Europe in terms of its role in society and level of intensification. In general, in CEE national agricultural production plays a much more important role in economy and society compared to WE. Also the proportion of the human population employed in agriculture is several times larger in CEE compared to WE. Majority of these results are the effect of political dividing of the continent after the Second World War, what made a situation called "an iron curtain" [31]. At the same time, CEE countries differ in the outcomes of past agricultural intensification: in some states, such as the Czech Republic and much of Slovakia, the communist "collectivist" agriculture created large monoculture fields, not unlike those in Northwestern Europe [31]. In many CEE countries though, small family farms have retained smaller field sizes and farming methods remain as they were decades ago. In Poland, nearly half of the > 2.5 million farms are still smaller than 2 ha, and this field mosaic is enriched by a dense network of seminatural field margins [32]. In some others, such as Hungary or Romania, mixed systems with intensive agriculture exist side-by-side with traditional farming in remote areas. Generally, small farms (<5 ha) are much more abundant in CEE than in WE, for example, there are over 50 times more such farms in Romania than in the UK and nearly 20 times more in Poland than in Germany.

The EU has so far failed to stop biodiversity loss in farmland [33, 34]. Although the action plan started few decades ago, the results are far away from the satisfactory situation. Currently published papers (a reviewed by ([27]) suggests that the management solutions developed mainly in Western Europe should not be used as a blanket prescription for the whole continent. The concept HNV farmland was not directionally tested in this particular context, but already existing results widely confirm this point of view [13, 35, 36].

1.3   Conservation Tools: From the Protected Areas Approach and Nature2000 Network to the High Nature Value Areas in Europe

Habitats and landscapes of Europe are increasingly impacted by the intensification of agriculture [21, 29, 37], the development of infrastructure, and increase of built up areas. The impact of human development on biodiversity is, however, not uniform across the continent. The conservation of biodiversity is always dependent on how humans perceive (model) natural systems and (based on this and the economical possibilities) how the management priorities are established. With the massive habitat, species and population loss and landscape homogenization, the most obvious reaction was to preserve areas where the biodiversity is still well represented and establish (threatened) species and habitat lists, on the base new protected areas are to be delimited. Actually there are several conventions and directives that guide biological conservation in the European Union. Besides many national level protected area types, the Natura 2000 sites represent the European Union's network of protected areas, delimited using the habitat and species lists established by the annexes of Habitats and Birds Directives. Although the annexes were updated with the addition of new member states (because the addition of new habitats and species with the new areas), the basics of protected area designations remained the same across the European Union. Specifically with regard to nature conservation, the European policy strengthened the implementation of a rational development strategy by influencing the Member States to adopt international commitments such as the Convention on Biological Diversity, and through the expansion of nature conservation areas. Among the EU directives promoting nature conservation, the most important provisions were the Birds Directive and the Habitats Directive. Implementation of these two directives subsequently gave rise to a new form of nature conservation - the Natura 2000 European Ecological Network (Fig. 1.1). At the regional level of the EU, the general principles and the implementation of the nature conservation policy are complex and governed in a top-down manner [38]. Such approach is inherently at risk of being introduced locally with a low level of effectiveness and adaptability, what makes a lot of troubles with local societies understanding of nature protection [38] and hence, current mechanisms of nature protection (mainly biodiversity) in the EU need to be complemented with effective bottom-up initiatives in addition to new means of top-down approaches.

The ecological network Natura 2000 differs considerably from the previous traditional protection system in that it aims at halting the biodiversity loss and maintaining or reconstructing the favorable nature conservation status by protecting

Legend

SITETYPE

Bird Directive sites

Habitat Directive sites

Sites belonging both Directives

Fig. 1.1 Distribution of the network of sites Natura2000 in Europe, described as the site type category. (Copyright holder: European Environment Agency (EEA))

natural habitat types, besides protection of floral and faunal species that are unique in the European continent [38, 39]. However, even under the best protection regime only the part of the environment is protected. In case of farmland, it make generates very interesting, both from theoretical and practical point of view, consequences.

1.3.1 A Concept of Protected Areas Application to Successful Conservation

Therefore, a conceptual model of biodiversity conservation (i.e. nature protected areas as HNV and Natura 2000, species lists with various protection status) was spatially transferred over Europe, including the Eastern part of the continent, where the biodiversity patterns and species richness are largely different than in those countries, where this model of protect biodiversity appeared. Complex mathematical analyses are proposed to assess the efficiency of protected area networks and proposals were made to increase it (i.e. regarding spatial distribution, size, isolation) [40]. However, even protected areas are exposed to various stochastic and deterministic (intrinsic and extrinsic) factors. All these may induce population and biodiversity decline, addition of new (invasive) species and changes in the community structure [41] and ultimately may compromise the specific objective of the site (to protect biodiversity).

1.3.2   Theories for Fragmented Space

The theory of island biogeography (TIB) [42] is considered the most influential root of the actual models of spatial ecology [43]. TIB explains the distribution patterns of species and communities on islands, using variables such are the island area and distance from other islands and the mainland (continent). Island area influences the species number and the persistence of populations. Isolation from mainland may be decisive factor in colonization. Small islands that are distant from mainland may be species poor because the high rate of extinction and low rate of colonization. Large islands, closely situated to mainland may have species rich communities and lower extinction rates. Moreover, the biotic interactions from the islands (in relation with area and isolation) also may influence the colonization-extinction events. TIB was developed for terrestrial organisms using "binary" space: islands acting as habitats, and the sea representing an increasing risk of mortality [44].

Researchers become increasingly aware of the massive terrestrial habitat destruction and fragmentation due to the increasing and complexely interacting human impacts: land use intensification, increasingly dense infrastructural network, pollution and increasing urbanization. All these impacts result in the split of formerly "continuous" habitats in small, more or less isolated, island like patches [45]. Under these conditions, the conceptually simple TIB quickly gathered an increased popularity among terrestrial ecologists and conservation biologists [46, 47]. The proposals of nature reserve design based on TIB (i.e. [46]) represent one example on how TIB was proposed to be used in conservation biology. The conceptual base of TIB (i.e. the existence of habitats in the "non-habitat" sea) represented the basic idea for metapopulation theory and landscape ecology [45, 48].

Metapopulation theory examines the distribution and persistence of species considering parameters such are the patch connectivity, extinction / (re)colonization of the vacant habitats. The early model of Levins (1969) assumed an infinite number of identical, equally connected patches. The patches were in empty or occupied state and had equal transition probabilities. Later, the classical Levins model was relaxed, with limiting the number of patches (homogenous stochastic patch occupancy models) and later adding heterogeneity to within and between patch quality (heterogeneous stochastic patch occupancy models) [43, 47]. This later development leads to spatially realistic metapopulation theory, where the transition probability depends on the landscape attributes. "Landscape" in this sense means networks of dissimilar patches [43]. According to the degree of isolation (and implicitly movement frequency) between patches and the demographical conditions, many types of (meta) population networks were distinguished besides the classical (Levins) model: mainland – island metapopulations, patchy- and non-equilibrium metapopulations. These may not be "true" metapopulations (true meaning that the four conditions of metapopulations is simultaneously met (see details in [50]).

Landscape ecology focus on the importance of environmental conditions across spatial scale to determine the distribution of organisms. In its early form, it considered the space as the totality of "patches" of different quality. Landscape elements that positively influence the occurrence-distribution of the focal organism are regarded as 'habitat' otherwise they are considered "matrix" (i.e. non-habitat) [51]. The associations are, of course, species-specific and a landscape element that is habitat, favorable for one species may be 'matrix' for another [52]. The structure of such a patchy landscape is described by i) landscape composition (i.e., the type, size and number of patches) and ii) landscape configuration (i.e., the patches' spatial arrangement of the patches relative to each other). The term 'landscape complementation' refers to the provisioning of landscape patches [53]: a landscape with high complementation means that landscape elements with critical resources are available for the organisms whereas a low complementarily means that the lack one (or more) landscape elements may limit the organism distribution.

The meaning of concepts may be different in the two approaches (metapopulation- and landscape ecology). The term "connectivity" is a (habitat) patch attribute according to the metapopulation theory (i.e. [54]). Large and closely situated patches contribute more to connectivity than small isolated ones. Therefore, the large, closely situated protected area "patches" will increase the persistence of focal organisms. Landscape ecology (however) addresses the connectivity as an attribute of the whole landscape, distinguishing two types of connectivity: structural (based on landscape elements) and functional (based on movement behavior of organisms through the landscape and its consequences) [52]. A highly connected landscape will allow movement of organisms within habitats (patches) without increasing the risks of mortality associated with inters habitat (patch, protected area) movements. Corridors are sometimes considered as permeable matrix that allow organisms to pass them and may or may not contribute to natality. Therefore, it is better for an organism to be on a "corridor" when live the habitat (patch) than in the matrix. According to this view, the areas between the protected areas should be managed in a way to allow movements of individuals across them. However, the large majority of the landscape studies still approach the landscape as the totality of patches (fragments) (review by [48], [55]).

Theories and approaches mentioned above were developed and empirically tested largely in binary type of space: organisms are linked to habitat patches that have different sizes, shape, and quality and at various distances from each other. This is understandable because island like systems are clearly delimitable and therefore useful for the development of strong theoretical framework [50]. Representation of space using patches (occasionally) connected with corridors is easy (see for example a map), and has a relatively low degree of complexity that makes them easy to understand [51]. Being such, these models are attractive especially for conservation biologists, who apply them for the understanding of the spatial distribution of organisms, proposing management interventions and protected areas (many studies). Indeed, the persistence of some species in highly fragmented landscapes, such are many intensively used agricultural landscapes from Western Europe largely depend on interconnected protected area networks [56].

1.3.3 Applying Ecological Terminology for Agricultural Landscapes of Whole Europe?

Biological conservation is largely built on "unclear certainties" reflected by terms like "habitat", "population", "ecosystem" and various terms denoting spatial and temporal dynamic of populations and communities such are: "turnover", "colonization", "extinction", "immigration", "emigration", "source-sink". All these are, nevertheless, suited for making complex spatial models. Below we show that many of these terms become vague, even questionable when applied to some species, mainly in Eastern and Central European landscapes [26, 48]. Therefore more attention should be pay when applying them.

The concept of habitat is fundamental in ecology and management for biodiversity conservation. We should understand the habitat requirement of organisms, and also, the spatial extent of that (particular?) habitat to predict organism distribution and the effect(s) of habitat loss. The term habitat is used in various ways by ecologists and conservationists. In a recent review, Mitchell (2005) distinguishes three ecological meanings of habitat: (1) the space used by organism, (2) the range of physico-chemical characteristics of the environment that determine/limit organism's distribution (i.e. the abiotic subset of the Hutchinsonian multidimensional niche) and (3) the community concept of habitat, based on complexes of plants and animals (grassland habitat, forest habitat etc.). Shreeve and Dennis, (2004) argue that the resources define the habitat and population structure: when the resources are scarce and patchily distributed in the landscape, the focal species may show metapopulation structure. However, when resources are diffuse in the landscape, such a definition is not applicable. Fischer and Lindenmayer, (2007) define habitat as "the range of environments suitable for a particular species". The definition (1) had no clear information regarding the factors controlling the distribution of organisms [57], definition (2) is to narrow (i.e. organisms distribution may be significantly influenced by biotic elements as well) and the habitat "hypothesis" under definition (3) should be tested (i.e. is the grassland [the only] habitat for species A?).

The various meanings of habitat concept, together with a high diversity of life histories of organisms (see below) may lead to increasing confusions in biodiversity conservation of Eastern European landscapes. For example, some landscape elements (such are wood pastures) are not protected as "habitats" under the Habitats Directive, therefore, not eligible for designation of SCI's (one kind of Natura 2000 sites). However, because of the species rich fauna (species from Annex 2 of Habitat Directive and the Birds Directive) virtually entire landscapes (with elements ranging from grassland, wetland, arable land, forest, wood pastures and even buildings and villages of traditional rural communities) may be used by one or more species (especially bats, birds) and therefore need to be protected.

In context of HNV farmland, Eastern European landscapes are characterized by high level of species richness. Species protected under international (i.e. European Union) legislation may be widely distributed. Therefore, apart to some species that are linked to patchily distributed habitats, delimiting population borders for species is difficult or impossible. Most probably, contour like approach (sensu Manning et al., (2004) would reveal some distributional peaks of various species but still, it is a dangerous adventure to say for some areas that they are inhospitable (i.e. "matrix"). This is because the population density peaks (i.e. high density of individuals in a certain place) are often not fixed. In these conditions it is difficult to categorize a land as "matrix" for these species, also, the source habitats-populations may be changing (see [61]) for more reflections about the difficulties of studying metapopulations and the spatial structure of populations in very patch nature-rich areas).

The above presented aspects highlight the fuzzy being of habitats and populations, and this may characterize wide parts of Eastern European countries. If we consider space as a complex and connected (temporarily, or more permanently) network of habitats of which quality is ever changing [26, 61], concepts such are "colonization", "extinction", "immigration", "emigration" may be not useful in understanding spatio-temporal dynamic of populations and communities in these landscapes. It is visualized in the Fig. 1.2.

As Fischer and Lindenmayer [51] noticed, the success of biological conservation is highly dependent on the way, how we conceptualize the ecological space and the distribution of various species and communities on it. Several organisms that have now limited distribution in some Western European countries are still widely present in Eastern Europe [26, 27]. For this reason, the habitat-matrix approach, and therefore spatial models such are area and isolation paradigm and (fragmentation) landscape models, may hold for some organisms that have patchy distributions in intensively used, fragmented landscapes, but may not be applicable for the same organisms found in traditionally managed areas.

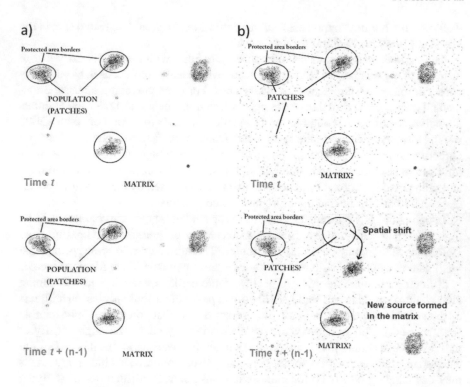

Fig. 1.2 A schematic representation of the possible distribution of an organism in a landscape with obvious habitat – matrix character (**a**) and in a landscape where the habitat – matrix character is not obvious. In the case of (**a**) the population borders of the organism are well delimitable and the various spatial models that incorporate turnover, extinction, colonization, connectivity, landscape composition and configuration may work. Moreover, protecting species and communities in this landscape can be realized by delimiting areas borders. Contrary, all these concepts and measures for protecting biodiversity may not work in the landscape (**b**). The drawing represents a snapshot for two moments (time t and time t + [n – 1]). The spatial arrangements of populations from (**a**) cannot change because the matrix is inhospitable (thus, the mortality risk is high). Here local extinctions may occur (small populations may be more prone), followed by recolonisation. In the case of landscape (**b**), the "matrix" may become important habitat and source of individuals for the entire landscape. Here spatial shifts of high population densities may occur (populations may even "leave" the borders of protected areas) and new source populations may appear in the matrix. Protecting biodiversity in landscape (**b**) by delimiting (and fixing) areas with high density, and ignoring the potentially source character of the "matrix" can be detrimental

The graph was conceptually prepared thanks to courtesy of Dr. T. Hartel during a work with the paper [26]

1.4   HNV Farming Definition

In an effort to protect farmland biodiversity, several studies were performed to assess the quality of agricultural landscapes, and in defining high nature value (HNV) farmland [62, 63]. Originally, the term HNV was introduced by Baldock

[64] and Beaufoy [65]. Despite the difficulty of defining HNV farming, the concept has continued evolving. More recently Andersen [66] proposed a conceptual definition for HNV farmland as 'those areas in Europe where agriculture is the dominant land use and where agriculture supports or is associated with either a high species and habitat diversity or the presence of species of European conservation concern or both'. This was subsequently followed by a mapping approach at EU level [2]. HNV farmland (represented by a binary indicator) is defined as one of the following three types [66]: farmland with a high proportion of semi-natural vegetation, farmland dominated by low intensity agriculture or a mosaic of semi-natural and cultivated land and small-scale features, and farmland supporting rare species or a high proportion of European or World populations. The first two types are identified on the basis of land cover data (CORINE database) and agronomic farm-level data (in particular the Farm Accounting Data Network), and the third type of HNV can only be identified by means of monitoring species occurrence.

Summarized, the HNV is an indicator referred to farming systems which include semi-natural habitats, low intensity farming and diverse, small-scale mosaics of land-use types [35] (Figs. 1.3 and 1.4). The high nature values (HNV) farming is present in all European countries, with a diversity of types and extensions. The need for measures to prevent the loss and monitoring in real time the quality of HNV areas and to mitigate the loss of biodiversity is widely recognized, but requires urgent attention. One of the reasons is due to the deep differences between agro ecosystems of Central-Eastern and Western Europe, differences that can explain some lack of achievements of several conservation policies, and later also in practice. Several recent studies showed how agri-environmental measures adopted in the countries of European Community may be ineffective in order to guarantee the stop on biodiversity in some intensive farmland [26]. Therefore, more studies

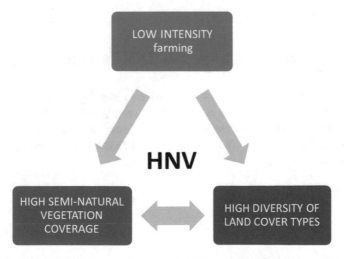

Fig. 1.3 Schematic representation of the concept used to define the high nature value farming in Europe, as provided in [63]

Fig. 1.4 Example of agricultural landscape from Central Italy mirroring the main characteristics of high nature farmlands, with the typical spatial disposition of land use features as a chessboard (Photo: F. Pruscini)

focusing strategies to increase the effectiveness of agri-environmental measures for conservation are necessaries.

1.5 Approaches to Characterize HNV Farming

Many different approaches were used in order to identify HNV farming zones in Europe, greatest combining farm systems, land cover/land use, or indicator species [67]. However, are the same elements that previously were elaborate by [66] and later developed by [68]. Follows the characteristics associated with these methods are described:

1.5.1 Land Cover Approach

The agricultural habitats with potential HNV are analyzed based on Corine land cover classes (LCCs) (Table 1.1) [68]. The CORINE classification has a low level

Table 1.1 Corine land cover classes (LCCs) considered potentially related to agricultural land

Code	Land cover class
2.1.1	Non-irrigated arable land
2.1.2	Permanently irrigated land
2.1.3	Rice fields
2.2.1	Vineyards
2.2.2	Fruit trees and berry plantation
2.2.3	Olive groves
2.3.1	Pastures
2.4.1	Annual crops associated with permanent crops
2.4.2	Complex cultivation patterns
2.4.3	Land principally occupied by agriculture with significant natural vegetation
2.4.4	Agro-forestry areas
3.2.1	Natural grasslands
3.2.2	Moors and heath lands
3.2.3	Sclerophyllous vegetation
3.2.4	Transitional woodland-scrub
3.3.3	Sparsely vegetated areas
4.1.1	Inland marshes
4.1.2	Peat bogs
4.2.1	Salt marshes

Source: [69]

of detail, the minimum resolution is 25 hectares units, which does not allow the distinction between intensively used grassland or extensive. Therefore, to refine the selection criteria, it is necessary to consider the environmental area and add additional information such as the land elevation, soil quality, slope inclination and distribution of relevant species [68].

1.5.2 Farm System Approach

These type of approaches are based on diverse features as for example, low levels on agrochemicals, farm size and high crop diversity [67] (Table 1.2). If combined with land cover approach, is possible to obtain maps of high detail resolution [63].

The farm systems methods use a features classification on production, inputs and management that are distinguished as following [63, 69]:

1. High Nature Value cropping systems, are characterized for their low intensity arable systems. It can also include livestock without being a main source of income;
2. High Nature Value permanent crop systems, are characterized by low intensity olives and other permanent crop systems;

Table 1.2 Definition of high nature value farming types

	Western Europe and Scandinavia	Southern Europe
HNV cropping systems	Input cost < 40 Euro/ha	Fallow systems: > 20.5 % of UAA in fallow and input cost < 40 Euro/ha Dryland systems: Not fallow systems and < 10 % of UAA irrigated and input cost< 40 Euro ha
HNV permanent crops	No data	Systems with grazing livestock: Input cost on crop protection < 10 Euro/ha and no irrigation and ≥ 5 grazing livestock units. Systems without grazing livestock: Input cost on crop protection < 10 Euro/ha and no irrigation and < 5 grazing livestock units
HNV off-farm grazing systems	≥ 150 grazing days outside UAA	≥ 150 grazing days outside UAA
HNV permanent grassland systems	Rough grassland systems: rough grassland ≥ 66 % of UAA and stocking density< 0.3 livestock units/ha Permanent grassland systems: rough grassland < 66 % of UAA and stocking density< 1.0 livestock units/ha	Stocking density < 0.2 livestock units/ha
HNV arable grazing livestock systems	Input cost < 40 Euro/ha	Input cost < 40 Euro/ha and ((≥ 20 % of UAA in fallow) or (0 % of UAA irrigated))
HNV othersystems	Input cost < 40 Euro/ha	Input cost < 40 Euro/ha and ((≥ 20 % of UAA in fallow) or (no irrigation))

Source: [69]
Note: *UAA* utilized agricultural area
HNV high nature value

3. High Nature Value off-farm grazing systems, these typologies are described by to have cattle and sheep that to graze outside the farm;
4. High Nature Value permanent grassland systems; the main forage resource for livestock is grass from permanent or rough grassland;
5. High Nature Value arable grazing livestock systems, the main forage resource for livestock is arable crops;
6. High Nature Value other systems, with mainly low intensity pigs or poultry systems.

Table 1.3 Categories of species of European conservation concern (SPEC)

Category 1	Species of global conservation concern because they are classed as globally threatened, conservation dependent or data deficient
Category 2	Species whose global populations are concentrated in Europe (i.e. more than 50 % of their global population or range in Europe) and which have an unfavorable conservation status in Europe
Category 3	Species whose global populations are not concentrated in Europe, but which have an unfavorable conservation status in Europe
Category 4	Species whose global populations are concentrated in Europe (i.e. species with more than 50 % of their global population or range in Europe) but which have a favorable conservation status in Europe

Source: [72]

1.5.3   Species Approach

Another approach is the HNV farmland indicator based on a landscape elements and indicator plant and birds species [70]. Through an assessment of the conservation status of all Europe's birds was performed a list of species of European conservation concern (SPECs) (Table 1.3). In this way, SPEC designation highlights the bird species of especially high conservation concern [71, 72]

Currently, the largest HNV areas remain mainly in eastern and southern Europe (Fig. 1.5), as well as in mountain areas, and contain habitat types such as semi-natural grasslands, dehesas and montados (open, wooded pastures in Spain and Portugal, respectively), steppe grasslands, permanent crops (such as fruit and nut orchards and olive groves), and arable crops in dryland areas where naturally regeneration through one to three year fallow is used to help rebuild soil nutrients before the next non-irrigated crop is planted (see a detailed treatment in [2], [66]), Fig. 1.6).

1.6   The HNV as Support for Biodiversity and Public Goods

Organic farming and low intensity agriculture can mitigate the negative effects of agricultural intensification on biodiversity [73], because biodiversity generally decreases with increased fertilizers and pesticide inputs, large use of machinery and intensive in productivity. Concerns about current Common Agricultural Policies (CAP) highlighted the negative trend in biodiversity, entrusting each European country with responsibility to tackle biodiversity loss [15, 64, 74, 75], encouraging farmers to conserve biodiversity through the maintenance of extensive farming systems and preservation of semi-natural landscape features [76, 77].

In fact, HNV farmland comprises hotspots of biological diversity in rural areas, corresponding ecologically to more functional heterogeneity and therefore high values of biodiversity and species abundance [21, 76, 78–82]. The main reasons are

Fig. 1.5 Green areas show likelihood to contain primarily HNV land, on the basis of a stratified selection of CORINE land cover 2006 classes per country and environmental zone, and national biodiversity data when available. The values in the map are a proxy for the proportion of HNV in each 1 km² cell, 2012 update (Copyright holder: European Environment Agency (EEA))

related to the effects of natural and semi-natural features, that are characteristics of HNV farming, increasing the landscape heterogeneity and increasing niche availability for many animal species [79, 83] (Fig. 1.7). For these reasons, recently has been proposed that maintain or increase the proportion of semi-natural landscape elements and areas with low management intensity can be a beneficial practice for species conservation [11, 84, 85]. Accordingly, the European Union's conservation policy adopted an indicator to identify and monitor the changes in high nature value (HNV) farmland [65, 66].

Among these features of farmlands were identified mainly the hedgerows, uncultivated areas, shrubs and isolated trees [13, 86]. The presence of scattered trees constitutes a key habitat for many species, playing an essential role as keystone structures with strong implications for conservation [87]. Also the

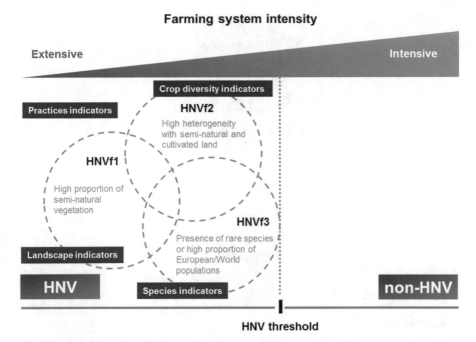

Fig. 1.6 Schematic representation of the concept used to define the high nature value farming types in Europe, as provided in [2]

marginal vegetation and hedgerows can imply an increase in the richness or abundance of species in agricultural landscapes [88]. In the same way, the presence of non-cropped areas or uncultivated patches can increase the biodiversity of bird communities in intensively used farmlands [89]. The Fig. 1.8 show some of the most typical features providing landscape heterogeneity in the HNV farmlands. Furthermore, recent studies conducted in France, investigated the relationship between measures of bird biodiversity and HNV farmlands. There, the authors found differences on farmland bird species abundances and composition of bird communities (measured by the community specialisation index, CSI) in relationship to the HNV farmland and not-HNV farmland. The same research had shown how HNV farmland plays a role in supporting the specialist bird species [90], offering clearly a potential tool to contrast the negative effects of biotic homogenization of bird communities [91, 92]. So, the effects of HNV farming practice on biodiversity are many, and the effects can be assumed at the level of taxonomic diversity as well as the level of functional diversity of animal communities (Fig. 1.7).

However, the HNV farmlands play a role also on human wellbeing, providing numerous direct and indirect benefits. Because the HNV landscapes are often agricultural landscapes, since HNV farmland is mostly the product of traditional practices, many time ancient ones. This is particularly true in the case of agricultural areas in Europe, where humans dominated land use for centuries, and

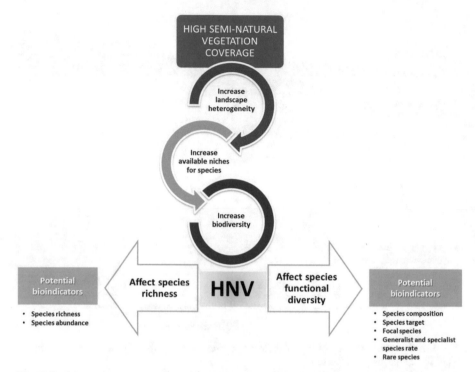

Fig. 1.7 Schematic representation of the main relationships between high nature farming characteristics as support for biodiversity, and potential bioindicator useful to detect the changes in HNV farmlands

sometimes millennia [65]. Then, the value that society gives to HNV farmland is not only anchored to the biodiversity conservation [93]. And this is one of the reasons because policy instruments have to keep safe this landscapes: because the HNV farming systems represent a socio-ecological systems [67].

1.7   This Book in Few Words

The aim of this book is to help to establish a simple framework to identify and use bird species as bioindicator for high nature values (HNV) farmlands. We focus our studies on suitable methods for monitoring the HNV areas, presenting the results of some case studies. Was proposed to integrate ecosystems assessment, geographical information systems (GIS) and strategies to conservation of local biodiversity. The proposed framework is focused on the use of species distribution models (SDMs) in order to explore the importance of each characteristic of HNV farmlands. Additionally, were examined the relationships among bird species richness, land use diversity and landscape metrics at a local scale in the farmlands.

Fig. 1.8 Examples of the most typical vegetation features providing landscape heterogeneity in HNV farmlands (**a**): tree hedgerows, (**b**): isolated trees, (**c**): shrubs and uncultivated patches, (**d**): marginal shrubs and hedgerows used to separate croplands (Photo: A. Wuczyński, F. Morelli, Y. Benedetti)

An innovative framework is suggested to develop models, using the HNV farmlands and the presence or abundance of birds. Finally, performing traditional SDMs on the bird species selected during the first step, is possible to study which characteristics of farmlands are driven the distribution of these species "indicators". These results encourage the possibility to identify accurately the HNV farmland by the presence of just few common bird species, and then the possibility to monitor it with a this cost-effective instrument, even linking common people to programs based on citizen science.

Furthermore, we discussed how the modern agricultural techniques, which simplify the structural complexity of farmland, are likely to exclude many passerine species. Thus incentives to maintain small scale heterogeneity in traditionally managed farmland will be critical for maintaining their rich passerine bird communities. Once such groups of species have been identified for different regions, these species can provide a fast and inexpensive tool for assessing the ecological value of farmland. This simple approach, can constitute a framework useful to be adopted in ecological planning at multi-spatial scale, in order to monitor changes in HNV areas related to particular features that characterize this environments (hedgerows, uncultivated patches, marginal vegetation), supporting also the biodiversity.

In this book are provided several instruments and information valuable in order to study the health of HNV areas in Europe.

Was conducted a systematic review with a comprehensive research in peer-reviewed journals in order to quantify the scientific interest on the concept of high nature value (HNV) in Europe, from the initial introduction of the concept in the scientific arena to nowadays. The reviewed studies provide an overall overview of the scientific findings of HNV studies previously conducted in Europe, country by country, and year per year. This information can help to identify the gaps in the current HNV knowledge and thus to highlight the future priorities for research.

Because all semi-natural features which characterize the HNV farmlands can be studied at the level of landscape spatial configuration by a geographical information system (GIS), we provided a summarized description of the concept and types of landscape metrics (LM) and its potential use for identify HNV based on land-use map characteristics.

A briefly description of some analytical and statistical techniques, commonly used in species distribution models and other ecological applications is provided. We provided also three detailed study case from different European countries: Italy, Portugal and Poland, with clear schemes to illustrate the main findings and to be used as guidelines for other studies based in bird species and HNV areas.

All the illustrated methodologies, can constitute an useful tool for track trends in HNV farmlands over time and for ecological restoration planning at the local scale, in the slipstream of European agricultural policy. An important key-word, considering that one of the main aims of the Common Agricultural Policy (CAP) is the reduction in the negative trends in biodiversity, entrusting each European country with the specific task of biodiversity conservation within its own territory. Then, the main benefits of the book could be taken by the many stakeholders, currently involved in the ecological assessment and conservation of widespread environments as farmlands are in Europe.

References

1. Balmford A, Bennun L, Brink B, Cooper D, Côté IM, Crane P, et al. The convention on biological diversity's 2010 target. Himal J Sci. 2005;3(5):43–5.
2. Lomba A, Guerra C, Alonso J, Honrado JP, Jongman RHG, McCracken D. Mapping and monitoring high nature value farmlands: challenges in European landscapes. J Environ Manage [Internet]. Elsevier Ltd; 2014;143:140–50. Available from: http://dx.doi.org/10.1016/j.jenvman.2014.04.029
3. Chamberlain DE, Fuller RJ, Bunce RGH, Duckworth JC, Shrubb M. Changes in the abundance of farmland birds in relation to the timing of agricultural intensification in England and Wales. J Appl Ecol [Internet]. 2000 Oct [cited 2013 Aug 6];37(5):771–88. Available from: http://doi.wiley.com/10.1046/j.1365-2664.2000.00548.x
4. Donald PF, Green RE, Heath MF. Agricultural intensification and the collapse of Europe's farmland bird populations. Proc R Soc London B - Biol Sci [Internet]. 2001 Jan 7 [cited 2014 Mar 21];268(1462):25–9. Available from: http://www.pubmedcentral.nih.gov/articlerender.fcgi?artid=1087596&tool=pmcentrez&rendertype=abstract

5. Donald PF, Sanderson FJ, Burfield IJ, Van Bommel FPJ. Further evidence of continent-wide impacts of agricultural intensification on European farmland birds, 1990–2000. Agric Ecosyst Environ [Internet]. 2006 Sep [cited 2013 Aug 14];116(3–4):189–96. Available from: http://linkinghub.elsevier.com/retrieve/pii/S016788090600079X

6. Geiger F, Bengtsson J, Berendse F, Weisser WW, Emmerson M, Morales MB, et al. Persistent negative effects of pesticides on biodiversity and biological control potential on European farmland. Basic Appl Ecol [Internet]. 2010 Mar [cited 2014 Apr 29];11(2):97–105. Available from: http://www.sciencedirect.com/science/article/pii/S1439179109001388

7. Sanderson FJ, Kucharz M, Jobda M, Donald PF. Impacts of agricultural intensification and abandonment on farmland birds in Poland following EU accession. Agric Ecosyst Environ [Internet]. 2013 Mar [cited 2014 May 4];168:16–24. Available from: http://www.sciencedirect.com/science/article/pii/S0167880913000273

8. Stoate C, Báldi A, Beja P, Boatman ND, Herzon I, van Doorn A, et al. Ecological impacts of early 21st century agricultural change in Europe – a review. J Environ Manag. 2009;91(1):22–46.

9. Laaksonen T, Lehikoinen A. Population trends in boreal birds: continuing declines in agricultural, northern, and long-distance migrant species. Biol Conserv [Internet]. Elsevier Ltd; 2013 Dec [cited 2013 Oct 17];168:99–107. Available from: http://linkinghub.elsevier.com/retrieve/pii/S0006320713003169

10. Kleijn D, Rundlöf M, Scheper J, Smith HG, Tscharntke T. Does conservation on farmland contribute to halting the biodiversity decline. Trends Ecol Evol [Internet]. 2011 Jun [cited 2011 Jun 27];26(9):474–81. Available from: http://linkinghub.elsevier.com/retrieve/pii/S016953471100142X

11. Benton TG, Vickery J, Wilson JD. Farmland biodiversity: is habitat heterogeneity the key. Trends Ecol Evol [Internet]. 2003 Apr [cited 2011 Jul 16];18(4):182–8. Available from: http://linkinghub.elsevier.com/retrieve/pii/S0169534703000119

12. Perkins AJ, Whittingham MJ, Morris AJ, Bradbury RB. Use of field margins by foraging yellowhammers *Emberiza citrinella*. Agric Ecosyst Environ. 2002;93:413–20.

13. Morelli F. Relative importance of marginal vegetation (shrubs, hedgerows, isolated trees) surrogate of HNV farmland for bird species distribution in Central Italy. Ecol Eng [Internet]. Elsevier B.V.; 2013 Aug [cited 2013 May 16];57:261–6. Available from: http://linkinghub.elsevier.com/retrieve/pii/S0925857413001614

14. Ceresa F, Bogliani G, Pedrini P, Brambilla M. The importance of key marginal habitat features for birds in farmland: an assessment of habitat preferences of red-backed shrikes *Lanius collurio* in the Italian Alps. Bird Study [Internet]. BTO 2012 Aug 13 [cited 2015 Feb 15];59(3):327–34. Available from: http://www.tandfonline.com/doi/abs/10.1080/00063657.2012.676623#.VOSmWPmG_X4

15. Bennett AF, Radford JQ, Haslem A. Properties of land mosaics: implications for nature conservation in agricultural environments. Biol Conserv [Internet]. 2006 Nov [cited 2012 Oct 4];133(2):250–64. Available from: http://linkinghub.elsevier.com/retrieve/pii/S0006320706002850

16. Reijnen R, Foppen R, Ter BC, Thissen J. The effects of car traffic on breeding bird populations in woodland. III. Reduction of density in relation to the proximity of main roads. J Appl Ecol. 2008;32(1):187–202.

17. Wierzcholska S, Dajdok Z, Wuczynski A. Do bryophytes reflect the diversity of vascular plants and birds in marginal habitats. Scripra Fac Rerum Nat Univ Ostrav [Internet]. 2008;186:194–200. Available from: http://www.researchgate.net/publication/234678404_Do_bryophytes_reflect_the_diversity_of_vascular_plants_and_birds_in_marginal_habitats

18. Cardinale BJ, Duffy JE, Gonzalez A, Hooper DU, Perrings C, Venail P, et al. Biodiversity loss and its impact on humanity. Nature [Internet]. Nature Publishing Group; 2012 Jun6 [cited 2016 Jul 16];486(7401):59–67. Available from: http://www.nature.com/doifinder/10.1038/nature11148

19. Newbold T, Hudson LN, Arnell AP, Contu S,Palma A De, Ferrier S, et al. Has land use pushed terrestrial biodiversity beyond the planetary boundary? A global assessment. Science (80-). American Association for the Advancement of Science; 2016;353(6296):45–50.

20. Tilman D, Isbell F, Cowles JM. Biodiversity and Ecosystem Functioning. Annu Rev Ecol Evol Syst [Internet]. Annual Reviews 2014 Nov 23 [cited 2015 Dec 29];45(1):471–93. Available from: http://www.annualreviews.org/doi/pdf/10.1146/annurev-ecolsys-120213-091917

21. Kleijn D, Kohler F, Báldi A, Batáry P, Concepción ED, Clough Y, et al. On the relationship between farmland biodiversity and land-use intensity in Europe. Proc R Soc London B - Biol Sci [Internet]. 2009 Mar 7 [cited 2011 Jul 16];276:903–9. Available from: http://www.pubmedcentral.nih.gov/articlerender.fcgi?artid=2664376&tool=pmcentrez&rendertype=abstract

22. Daily G. Nature's services: societal dependence on natural ecosystems. Washington, DC: Island Press; 1997. 412 p.

23. Ricketts TH, Daily GC, Ehrlich PR, Michener CD. Economic value of tropical forest to coffee production. Proc Natl Acad Sci U S A [Internet]. National Academy of Sciences; 2004 Aug 24 [cited 2016 Jul 11];101(34):12579–82. Available from: http://www.ncbi.nlm.nih.gov/pubmed/15306689

24. Thies C, Roschewitz I, Tscharntke T. The landscape context of cereal aphid-parasitoid interactions. Proc Biol Sci [Internet]. The Royal Society; 2005Jan 22 [cited 2016 Jul 11];272 (1559):203–10. Available from: http://www.ncbi.nlm.nih.gov/pubmed/15695212

25. Blitzer EJ, Gibbs J, Park MG, Danforth BN. Pollination services for apple are dependent on diverse wild bee communities. Agric Ecosyst Environ [Internet]. 2016 Apr [cited 2016 Jul 11];221:1–7. Available from: http://linkinghub.elsevier.com/retrieve/pii/S0167880916300020

26. Tryjanowski P, Hartel T, Báldi A, Szymański P, Tobółka M, Herzon I, et al. Conservation of farmland birds faces different challenges in western and central-eastern Europe. Acta Ornithol [Internet]. 2011 Jun [cited 2013 Feb 19];46(1):1–12. Available from: http://www.bioone.org/doi/abs/10.3161/000164511X589857

27. Sutcliffe LME, Batáry P, Kormann U, Báldi A, Dicks LV, Herzon I, et al. Harnessing the biodiversity value of Central and Eastern European farmland. Divers Distrib. 2015;21 (6):722–30.

28. Bignal EM. Using an ecological understanding of farmland to reconcile nature conservation requirements, EU agricultural policy and word trade agreements. J Appl Ecol. 1998;35:949–54.

29. Tscharntke T, Klein AM, Kruess A, Steffan-Dewenter I, Thies C. Landscape perspectives on agricultural intensification and biodiversity – ecosystem service management. Ecol Lett. 2005;8:857–74.

30. Tryjanowski P, Dajdok Z, Kujawa K, Kałuski T, Mrówczyński M. Zagrożenia różnorodności biologicznej w krajobrazie rolniczym: czy badania wykonywane w Europie Zachodniej pozwalają na poprawną diagnozę w Polsce. Polish J Agron. 2011;7:113–9.

31. Reif J, Böhning-Gaese K, Flade M, Schwarz J, Schwager M. Population trends of birds across the iron curtain: brain matters. Biol Conserv. 2011;144(10):2524–33.

32. Wuczyński A, Kujawa K, Dajdok Z, Grzesiak W. Species richness and composition of bird communities in various field margins of Poland. Agric Ecosyst Environ [Internet]. 2011 Apr [cited 2013 Aug 19];141(1–2):202–9. Available from: http://www.sciencedirect.com/science/article/pii/S0167880911000831

33. Balmford A, Bennun L, Brink BT, Cooper D, Côte IM, Crane P, et al. Ecology: the convention on biological diversity's 2010 target. Science (80-). 2005;307:212–213.

34. Pedroli B, Van Doorn A, De Bust G, Paracchini ML, Washer D, Bunce F. Europe's living landscapes: essays exploring our identity in the countryside. Wageningen: KNNV Publishing; 2007.

35. Aue B, Diekötter T, Gottschalk TK, Wolters V, Hotes S. How high nature value (HNV) farmland is related to bird diversity in agro-ecosystems – towards a versatile tool for biodiversity monitoring and conservation planning. Agric Ecosyst Environ [Internet]. Elsevier B. V.; 2014;194:58–64. Available from: http://dx.doi.org/10.1016/j.agee.2014.04.012

36. Sutcliffe L, Akeroyd J, Page N, Popa R. Combining approaches to support high nature value Farmland in southern Transylvania. Romania Hacquetia. 2015;14(1):53–63.

37. Balmford A, Green RE, Scharlemann JPW. Sparing land for nature: exploring the potential impact of changes in agricultural yield on the area needed for crop production. Glob Chang Biol [Internet]. Blackwell Science Ltd; 2005 Oct [cited 2016 Jul 11];11(10):1594–605. Available from: http://doi.wiley.com/10.1111/j.1365-2486.2005.001035.x

38. Grodzińska-Jurczak M, Cent J. Expansion of nature conservation areas: problems with Natura 2000 implementation in Poland? Environ Manage [Internet]. 2011 Jan [cited 2014 Oct 26];47(1):11–27. Available from: http://www.pubmedcentral.nih.gov/articlerender.fcgi?artid=3016195&tool=pmcentrez&rendertype=abstract

39. Wuczyński A, Dajdok Z, Wierzcholska S, Kujawa K. Applying red lists to the evaluation of agricultural habitat: regular occurrence of threatened birds, vascular plants, and bryophytes in field margins of Poland. Biodivers Conserv [Internet]. Springer Netherlands; 2014 Apr 27 [cited 2016 Oct 12];23(4):999–1017. Available from: http://link.springer.com/10.1007/s10531-014-0649-y

40. Cabeza M, Moilanen A. Design of reserve networks and the persistence of biodiversity. Trends Ecol Evol. 2001;16(5):242–8.

41. Lovejoy TE. Climate change and biodiversity. New Delhi: The Energy and Resources Institute (TERI); 2006.

42. MacArthur RH, Wilson EO. The theory of island biogeography. Princeton: Princeton University Press; 1963. 203 p.

43. Hanski I, Ovaskinen O. Metapopulation theory for fragmented landscapes. Theor Popul Biol. 2003;64:119–27.

44. Whittaker R, Fernandez-Palacios JM. Island biogeography: ecology, evolution, and conservation. Second. New York: Oxford University Press; 2006. 416 p.

45. Fahrig L. Effects of habitat fragmentation on biodiversity. Annu Rev Ecol Evol Syst [Internet]. Annual Reviews 4139 El Camino Way, P.O. Box 10139, Palo Alto, CA 94303-0139, USA 2003 Nov [cited 2016 Oct 12];34(1):487–515. Available from: http://www.annualreviews.org/doi/10.1146/annurev.ecolsys.34.011802.132419

46. Diamond JM. The island dilemma: lessons of modern biogeographic studies for the design of natural reserves. Biol Conserv Elsevier. 1975;7(2):129–46.

47. Hanski I. Spatially realistic theory of metapopulation ecology. Naturwissenschaften [Internet]. 2001 Sep [cited 2016 Oct 12];88(9):372–81. Available from: http://www.ncbi.nlm.nih.gov/pubmed/11688412

48. Hartel T, Moga CI, Öllerer K, Sas I, Demeter L, Rusti D, et al. A proposal towards the incorporation of spatial heterogeneity into animal distribution studies in Romanian landscapes. North West J Zool. 2008;4:173–88.

49. Levins R. Some demographic and genetic consequences of environmental heterogeneity for biological control. Bull Entomol Soc Am. 1969;15(3):237–40.

50. Hanski I. Metapopulation dynamics. Nature. 1998;396:41–9.

51. Fischer J, Lindenmayer DB, Fazey I. Appreciating ecological complexity: habitat contours as a conceptual landscape model. Conserv Biol [Internet]. Blackwell Publishing Inc; 2004 Oct [cited 2016 Oct 12];18(5):1245–53. Available from: http://doi.wiley.com/10.1111/j.1523-1739.2004.00263.x

52. Tischendorf L, Fahrig L. How should we measure landscape connectivity? Landsc Ecol [Internet]. Kluwer Academic Publishers; 2000 [cited 2016 Oct 12];15(7):633–41. Available from: http://link.springer.com/10.1023/A:1008177324187

53. Dunning JB, Danielson BJ, Pulliam HR. Ecological processes that affect populations in complex landscapes. Oikos [Internet]. 1992 Oct [cited 2016 Oct 12];65(1):169–75. Available from: http://www.jstor.org/stable/3544901?origin=crossref

54. Moilanen A, Hanski I. Own the use of connectivity measures in spatial ecology. Oikos [Internet]. 2001;95(1):147–51. Available from: http://www.jstor.org/stable/3547357?seq=1#page_scan_tab_contents

55. Mazerolle MJ, Villard M-A. Patch characteristics and landscape context as predictors of species presence and abundance: a review. Ecoscience. 1999;6(1):117–24.

56. Evans D. Building the european union's natura 2000 network. Nat Conserv. 2012;1:11.
57. Mitchell SC. How useful is the concept of habitat? - a critique. Oikos [Internet]. Munksgaard International Publishers 2005 Sep [cited 2016 Oct 12];110(3):634–8. Available from: http://doi.wiley.com/10.1111/j.0030-1299.2005.13810.x
58. Shreeve T, Dennis RLH. Resources, habitats and metapopulations – whither reality? Oikos. 2004;106:404–8.
59. Fischer J, Lindenmayer DB. Landscape modification and habitat fragmentation: a synthesis. Glob Ecol Biogeogr [Internet]. Blackwell Publishing Ltd; 2007 May [cited 2016 Oct 10];16 (3):265–80. Available from: http://doi.wiley.com/10.1111/j.1466-8238.2007.00287.x
60. Manning AD, Lindenmayer DB, Nix HA. Continua and Umwelt: novel perspectives on viewing landscapes. Oikos [Internet]. Munksgaard; 2004 Mar [cited 2016 Oct 10];104 (3):621–8. Available from: http://doi.wiley.com/10.1111/j.0030-1299.2004.12813.x
61. Mikulcak F, Newig J, Milcu AI, Hartel T, Fischer J. Integrating rural development and biodiversity conservation in Central Romania. Environ Conserv. 2013;40(2):129–37.
62. Pointereau P, Doxa A, Coulon F, Jiguet F, Paracchini ML. Analysis of spatial and temporal variations of High Nature Value farmland and links with changes in bird populations: a study on France [Internet]. JRC Scientific and Technical Reports 2010. Available from: http://eln-fab.eu/uploads/Analysis of spatial and temporal variations of High Nature Value farmland and links with changes in bird populations a study on France.pdf
63. Bartel A. High Nature value farmland as an European evaluation indicator – definition, function and status quo. In: International workshop of the SALVERE-Project 2009. Agricultural Research and Education Centre gricultural Research and Education Centre Raumberg-Gumpenstein; 2009. p. 15–7.
64. Baldock D, Beaufoy G, Bennett G, Clark J. Nature conservation and new directions in the EC Common Agricultural Policy. London: Institute for European Environmental Policy (IEEP); 1993.
65. Beaufoy G, Baldock D, Clark J. The nature of farming: low intensity farming systems in nine European countries. London: Institute for European Environmental Policy; 1994.
66. Andersen E, Baldock D, Bennett H, Beaufoy G, Bignal EM, Brouwer F, et al. Developing a high nature value indicator. Report for the European Environment Agency [Internet]. Copenhagen; 2003. Available from: http://www.ieep.eu/assets/646/Developing_HNV_indicator.pdf
67. Strohbach MW, Kohler ML, Dauber J, Klimek S. High nature value farming: from indication to conservation. Ecol Indic [Internet]. Elsevier Ltd;2015;57:557–63. Available from: http://linkinghub.elsevier.com/retrieve/pii/S1470160X15002435
68. Paracchini ML, Petersen J, Hoogeveen Y, Bamps C, Burfield I, Swaay C Van. High nature value farmland in Europe – an estimate of the distribution patterns on the basis of land cover and biodiversity data. Institute for Environment and Sustainability Office for Official Publications of the European Communities Luxembourg;2008. 1–102 p.
69. Hoogeveen Y, Petersen J-EJ-E, Balazs K, Higuero I, EEA. High nature value farmland - Characteristics, trends and policy challenges. Vol. 1/2004, Office for Official Publications of the European Communities. 2004. 31 p.
70. Klimek S, Lohss G, Gabriel D. Modelling the spatial distribution of species-rich farmland to identify priority areas for conservation actions. Biol Conserv Elsevier Ltd. 2014;174:65–74.
71. Tucker G.M. and Heath M.F. Birds in Europe. Their conservation status. Birdlife Conserv Ser No 3 Birdlife Int Cambridge. 1994;
72. Tucker GM. Priorities for bird conservation in Europe: the importance of the farmed landscape. In: Pain D, PM W, editors. Farming and birds in Europe: the common agricultural policy and its implications for bird conservation. London: Academic press; 1997 .p. 79–116
73. Lüscher G, Ammari Y, Andriets A, Angelova S, Arndorfer M, Bailey D, et al. Farmland biodiversity and agricultural management on 237 farms in 13 European and two African regions. Ecology [Internet]. 2016 Jun [cited 2016 Jul 16];97(6):1625–1625 Available from: http://doi.wiley.com/10.1890/15-1985.1

74. de la Concha I. The common agricultural policy and the role of Rural Development Programmes in the conservation of steppe birds. In: Bota G, MB M, Mañosa S, Campro¬don J, editors. Ecology and conservation of steppe-land birds. Bellaterra/Barcelona: Lynx Edicions/Centre Tecnològic Forestal de Catalunya; 2005.

75. Oñate JJ. A reformed CAP? Opportunities and threats for the conservation of steppe-birds and the agri-environment. In: Bota G, Morales M B, Mañosa S, Camprodon J, editors. Ecology and conservation of steppe-land birds. Bellaterra/Barcelona: Lynx Edicions/Centre Tecnològic Forestal de Catalunya B; 2005.

76. Kleijn D, Sutherland WJ. How effective are European agri-environment schemes in conserving and promoting biodiversity? J Appl Ecol [Internet]. 2003 Dec [cited 2014 Apr 28];40 (6):947–69 Available from: http://doi.wiley.com/10.1111/j.1365-2664.2003.00868.x

77. Pain DJ, Pienkowski MW. Birds and farming in Europe: the Common Agricultural Policy and its implications for bird conservation. San Diego: Academic Press; 1997.

78. Hoffmann J, Greef JM. Mosaic indicators—theoretical approach for the development of indicators for species diversity in agricultural landscapes. Agric Ecosyst Environ [Internet]. 2003 Sep [cited 2011 Sep 8];98(1–3):387–94 Available from: http://linkinghub.elsevier.com/retrieve/pii/S0167880903000987

79. Fahrig L, Baudry J, Brotons L, Burel FG, Crist TO, Fuller RJ, et al. Functional landscape heterogeneity and animal biodiversity in agricultural landscapes. Ecol Lett [Internet]. 2011 Feb [cited 2014 Jan 23];14(2):101–12. Available from: http://www.ncbi.nlm.nih.gov/pubmed/21087380

80. Stirnemann IA, Ikin K, Gibbons P, Blanchard W, Lindenmayer DB. Measuring habitat heterogeneity reveals new insights into bird community composition. Oecologia [Internet]. 2014 Nov 7 [cited 2014 Nov 15];177:733–46. Available from: http://www.ncbi.nlm.nih.gov/pubmed/25376157

81. Flick T, Feagan S, Fahrig L. Effects of landscape structure on butterfly species richness and abundance in agricultural landscapes in eastern Ontario, Canada. Agric Ecosyst Environ [Internet]. 2012 Aug [cited 2014 Nov 14];156:123–33 Available from: http://www.sciencedirect.com/science/article/pii/S0167880912001843

82. Morelli F, Santolini R, Sisti D. Breeding habitat of red-backed shrike *Lanius collurio* on farmland hilly areas of Central Italy: is functional heterogeneity one important key? Ethol Ecol Evol. 2012;24(2):127–39.

83. Kisel Y, McInnes L, Toomey NH, Orme CDL. How diversification rates and diversity limits combine to create large-scale species-area relationships. Philos Trans R Soc London B - Biol Sci [Internet]. 2011 Sep 12 [cited 2011 Aug 14];366(1577):2514–25 Available from: http://www.pubmedcentral.nih.gov/articlerender.fcgi?artid=3138612&tool=pmcentrez&rendertype=abstract

84. Bengtsson J, Ahnström J, Weibull AC. The effects of organic agriculture on biodiversity and abundance: a meta-analysis. J Appl Ecol. 2005;42:261–9.

85. Billeter R, Liira J, Bailey D, Bugter R, Arens P, Augenstein I, et al. Indicators for biodiversity in agricultural landscapes: a pan-European study. J Appl Ecol [Internet]. 2007 Jul 23 [cited 2014 Jul 14];45(1):141–50. Available from: http://doi.wiley.com/10.1111/j.1365-2664.2007.01393.x

86. Morelli F, Girardello M. Buntings (Emberizidae) as indicators of HNV of farmlands: a case of study in Central Italy. Ethol Ecol Evol [Internet]. Taylor & Francis 2013 Nov 5 [cited 2013 Nov 11];26(4):405–12. Available from: http://www.tandfonline.com/doi/abs/10.1080/03949370.2013.852140

87. Manning AD, Fischer J, Lindenmayer DB. Scattered trees are keystone structures – Implications for conservation. Biol Conserv [Internet]. 2006 Oct [cited 2014 May 25];132(3):311–21. Available from: http://linkinghub.elsevier.com/retrieve/pii/S0006320706001807

88. Vickery JA, Feber RE, Fuller RJ. Arable field margins managed for biodiversity conservation: a review of food resource provision for farmland birds. Agric Ecosyst Environ [Internet]. 2009 Sep [cited 2016 Jul 24];133(1–2):1–13. Available from: http://linkinghub.elsevier.com/retrieve/pii/S0167880909001625

89. Chiron F, Filippi-Codaccioni O, Jiguet F, Devictor V. Effects of non-cropped landscape diversity on spatial dynamics of farmland birds in intensive farming systems. Biol Conserv [Internet]. 2010 Nov [cited 2016 Jul 24];143(11):2609–16. Available from: http://linkinghub.elsevier.com/retrieve/pii/S0006320710003095

90. Doxa A, Bas Y, Paracchini ML, Pointereau P, Terres J-M, Jiguet F. Low-intensity agriculture increases farmland bird abundances in France. J Appl Ecol [Internet]. 2010 Dec 21 [cited 2011 Jun 19];47(6):1348–56. Available from: http://doi.wiley.com/10.1111/j.1365-2664.2010.01869.x

91. Devictor V, Julliard R, Clavel J, Jiguet F, Lee A, Couvet D. Functional biotic homogenization of bird communities in disturbed landscapes. Glob Ecol Biogeogr [Internet]. 2008 Mar [cited 2011 Jul 19];17(2):252–61. Available from: http://doi.wiley.com/10.1111/j.1466-8238.2007.00364.x

92. Devictor V, Julliard R, Couvet D, Lee A, Jiguet F. Functional homogenization effect of urbanization on bird communities. Conserv Biol [Internet]. 2007 Jun [cited 2014 Jul 15];21(3):741–51. Available from: http://www.ncbi.nlm.nih.gov/pubmed/17531052

93. Paracchini M-L, Oppermann R. Public goods and ecosystems services delivered by HNV farmland. In: Oppermann R, Beaufoy G, Jones G, editors. High Nature Value Farming in Europe – 35 European Countries – experiences and perspectives. 2011; 2011. p. 446–50.

Chapter 2
Spread of the Concept of HNV Farmland in Europe: A Systematic Review

Yanina Benedetti

Abstract This chapter provides a systematic review regarding the spread of the High Nature Value (HNV) farmland concept in Europe, considering comprehensive research in peer-reviewed journals. Web of Sciences and Scopus were used as databases, focusing on several factors in order to explore the trends in HNV studies in different countries. The findings of this systematic review highlight that it is necessary to increase the number of HNV peer-reviewed articles in many European countries, expanding the focus to other taxa less studied than birds and plants, as well as considering other aspects of biodiversity—for instance, functional diversity or phylogenetic diversity—in order to achieve a more complete assessment of the conservation status of HNV farming areas.

Keywords HNV • Systematic review • Peer-reviewed articles • Biodiversity metrics • Taxon

2.1 Introduction

The concept of High Nature Value (HNV) farmland has been evolving over the last 30 years in Europe and several scientific articles related to HNV have been published from different countries of the European region.

A major aim of the European Union is to reduce the loss of biodiversity by 2020 [1]. According to this, the progress in HNV farming research has a key role in assessing the level of biodiversity and in monitoring the quality and status of HNV [2, 3]. Currently, HNV farming research—a discipline that is currently in development—involves several disciplines such as ecology, agronomy, veterinary science, economics, and social sciences. In synergy with this, EU and national agricultural and environmental policies, and large amounts of research and funding, are dedicated to biodiversity conservation approaches [4].

Y. Benedetti (✉)
Centro Naturalistico Sammarinese, via Valdes De Carli 21, 47893 Borgo Maggiore, San Marino
e-mail: ybenedetti73@gmail.com

© Springer International Publishing AG 2017   27
F. Morelli, P. Tryjanowski (eds.), *Birds as Useful Indicators of High Nature Value Farmlands*, DOI 10.1007/978-3-319-50284-7_2

In order to quantify the scientific interest in the HNV concept in Europe, we conducted a systematic review through comprehensive research in peer-reviewed journals. The reviewed studies can provide an overall overview of the characteristics of HNV studies conducted in Europe published to date by discussing the temporal trends, the geographical distribution of the research, the typologies of the articles, the main disciplines involved, the taxa used and, finally, in the case of studies on ecology, the main indices used to assess biodiversity. Second, based on our findings, this information enables us to identify the gaps in the current HNV knowledge and thus to highlight some important challenges for future research on HNV.

2.2   Methods

We conducted a comprehensive research in peer-reviewed journals. The databases used were the Web of Science™ (http://www.isiwebofknowledge.com) and Scopus (http://www.scopus.com), using the following search term combinations: (1) "HNV farmland", (2) "HNV farming", (3) "High Nature Value farmland", (4) "High Nature Value farming", (5) "HNV", and (6) "High Nature Value" in the TITLE and TOPIC sections.

We determined the exclusion and inclusion criteria a priori to screen the articles by title, abstract, and full text (Table 2.1). The selection of studies relevant for this systematic review was made in a two-stage process. First, relevance for the current study was initially assessed on the basis of study titles, abstracts, and keywords. Second, we analyzed the full text of all papers included in the final systematic review using a checklist to capture information about the papers in order to know different aspects of the basic information about the reviewed literature.

We excluded articles that were not peer-reviewed; articles in languages other than English; non-original papers such as theoretical reviews, book reviews, letters, editorials, summaries of conferences, or historical papers; and papers without abstracts. Some articles appeared in several academic databases. Duplicate papers in the databases were deleted.

The papers included in the review were examined and the following data were extracted in a database: (i) year of publication; (ii) country of research; (iii) taxon studied; (iv) subject area; (v) type of document (original article or review); and (vi) biodiversity measures used. These factors were examined to look for trends in HNV studies, as well as any gaps in how this topic was analyzed. If a study considered more than one parameter (e.g., country, taxon, etc.) we treated them as independent studies [5].

The several biodiversity metrics used in the articles were clustered in the following categories: taxonomic diversity, species abundance, species diversity/evenness, alpha/beta diversity, functional diversity, and others.

We calculated frequencies and percentages of the variables included in the review using SPSS 19 and MS Excel 2010 commercial software.

Table 2.1 List of inclusion
and exclusion criteria

Inclusion criteria
Papers must focus on High Nature Value
Peer-reviewed article (article or review)
Exclusion criteria
No peer-reviewed article
Publications without abstract
Paper not written in English

Fig. 2.1 Number of peer-
reviewed articles focused
on High Nature Value
(HNV) farmland by
typology

2.3   Results

The database search resulted in 286 articles. After screening titles, abstracts, and
full text, 207 articles were excluded. A total of 79 peer-reviewed studies of HNV
farmland in Europe were identified for the present study. Seventy-three articles
were original articles (92.4 %) and six were review articles (7.6 %) (Fig. 2.1).

HNV research has developed slowly. Since 2006 there has been a slowly
increasing number of scientific articles published; until 2014 studies in HNV
farmland did not exceed ten articles per year, doubling to 20 articles per year in
2015 (Fig. 2.2).

The subject area with the greatest number of articles was biodiversity and
conservation (54.4 %), followed by environmental sciences and ecology (17.7 %)
and agricultural and social sciences (7.6 %), the lower; fewer contributions came
from agricultural and biological sciences, veterinary sciences, public administration
and behavioral sciences, each of which accounted for 1–3 % of articles (Fig. 2.3).

Twenty-three EU countries published relevant articles on HNV farmland in
Europe. Most of the studies were conducted in Europe (20.25 %) (Fig. 2.4). Our

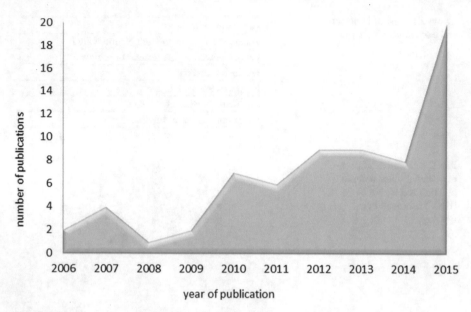

Fig. 2.2 Number of peer-reviewed articles focused on High Nature Value (HNV) farmland by year to 2015

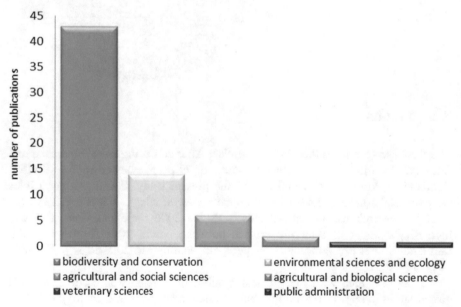

Fig. 2.3 Frequency of research areas of peer-reviewed articles focused on High Nature Value (HNV) farmland

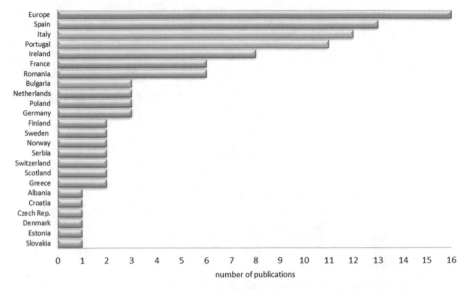

Fig. 2.4 Number of publications by country that have focused on High Nature Value (HNV) studies (*green bar*) and those considering the whole European community (*orange bar*)

analyses revealed that the most influential countries, in terms of numbers of peer-reviewed articles, are concentrated in the south of Europe: Spain (16.5 %), Italy (15 %), Portugal (14 %), and France (7.6 %) (Fig. 2.5). In particular, Italy has rapidly increased its research activities in recent years. In the North of Europe, Ireland has been the subject of studies on HNV farming in 10 % of cases. Romania stands out as the country in Central Eastern Europe with more studies on HNV (7.6 %). All other countries have very low relative shares (1–3 %) (Fig. 2.5).

In 47 of HNV farmland publications (59.6 %), different taxa were examined. The study results showed that the taxa most studied were birds (59.6 % of cases), plants (57.4 %), insects (34 %), and mammals (21.3 %). Few studies were found about amphibians (6.4 %), arachnid bryophytes and lichen (4.6 % for each one), and reptiles (1 %) (Fig. 2.6).

The biodiversity measures most frequently utilized to assess biodiversity status in HNV farmland studies were taxonomic diversity (66.7 %), followed by species diversity/evenness and alpha/beta diversity (10 % of studies). Other biodiversity metrics had very low relative inputs (Fig. 2.7).

2.4   Discussion

Systematic reviews are the best tool for a synthesis of primary results about topics of interest [6]. Pursuing this goal, this systematic review provides a synopsis of the most influential peer-reviewed scientific literature on HNV farmland in Europe.

Our results underscored that HNV farming systems have attracted attention from multiple discipline areas and they have employed heterogeneous approaches. Since

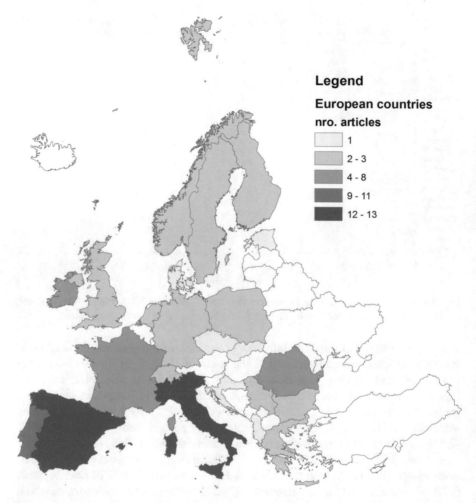

Fig. 2.5 Number of peer-reviewed articles focused on High Nature Value (HNV) farmland per country in Europe, using a *green*-scale gradient

the beginning of the 1990s, when the term was established by Baldock et al. [7], there has been a slow increase in the number of peer-reviewed scientific articles on HNV farming being published; the greatest slope was observed just in 2015, when the number of articles doubled to 20 articles for the year. Probably, this trend is due to an initial effort more focused on EU reports of guidelines in order to define, classify and assess HNV farming areas, rather than traditional research on the topic.

Considering the wide distribution of HNV farmland across Europe, the analysis of HNV studies for each EU country show there is a geographical bias. The largest number of HNV studies came from Spain and Italy, which were mostly devoted to biodiversity assessment of HNV areas, whereas in Eastern Europe, Romania was the country most dedicated to the study of HNV farmlands.

Fig. 2.6 Distribution of taxa studied in peer-reviewed articles focused on High Nature Value (HNV) farmland

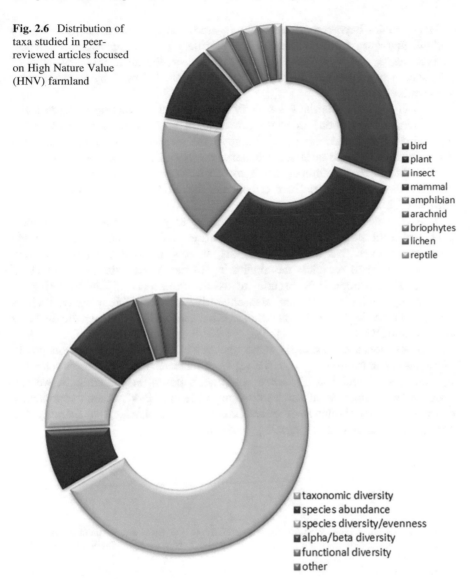

- bird
- plant
- insect
- mammal
- amphibian
- arachnid
- briophytes
- lichen
- reptile

- taxonomic diversity
- species abundance
- species diversity/evenness
- alpha/beta diversity
- functional diversity
- other

Fig. 2.7 Distribution of biodiversity metrics used in peer-reviewed articles focused on High Nature Value (HNV) farmland

In conservation management it is very important to identify surrogates that help to assess overall biodiversity [8]. In this sense, the bioindicators can be partial measures or estimator surrogates of biodiversity [9]. Our findings showed that taxon type, such as birds and plants, are the most useful bioindicators for studying biodiversity in HNV farming systems. This results are according to McKinney, who argues that birds, mammals, and plants are the best-studied taxa along urban–rural gradients [10].

HNV studies have come from different study disciplines, an interdisciplinary outlook that goes from biodiversity to veterinary science, agricultural science, social science, and public administration. However, the most prevalent areas of published articles are biodiversity and conservation, followed by environmental sciences and ecology.

In brief, these results show that HNV farming is a stimulating research arena where incipient directions are starting to crystallize. However, our analyses have highlighted that it is necessary to improve the synergies with related research disciplines, in order to understand better the HNV farming changes that can be attributed to external anthropogenic factors or to original natural processes, and to quantify the direct and indirect effects on the biodiversity supported by this particular kind of farming system.

It is also necessary to increase the number of peer-reviewed HNV articles published from the rest of the European countries and to expand them to include other less studied taxa such as mammals, insects, arachnids, amphibians, and reptiles in order to complete the monitoring of the conservation status of HNV farming. Furthermore, it is essential to assess the effects of HNV farming on biodiversity, taking into account also other dimensions of biodiversity, such as functional diversity [11, 12], in addition to other aspects like phylogenetic diversity and evolutionary distinctiveness [13].

In conclusion, these results show that research in HNV is a relatively young field of research, and the diversity of HNV research indicates that it possesses scientific dynamism and great potential for development. Based on our findings, we can expect a future increase in articles covering different HNV research approaches, contributing to the resolution of several issues in the monitoring, management, and ecological planning of HNV farming systems.

Bibliography

1. European Commission. Proposal for a regulation of the European Parliament and of the Council on support for rural development by the European Agricultural Fund for Rural Development (EAFRD); 2011.
2. Baldock D, Beaufoy G, Bennett G, Clark J. Nature conservation and new directions in the EC Common Agricultural Policy: report for the Ministry of Agriculture, Nature Management and Fisheries, the Netherlands. London: Institute for European Environmental Policy; 1993.
3. Bartolini F, Brunori G. Understanding linkages between common agricultural policy and High Nature Value (HNV) farmland provision: an empirical analysis in Tuscany region. Agric Food Econ. Berlin/Heidelberg: Springer; 2014 Dec;2(1):13
4. Farmer M, Cooper T, Baldock D, Tucker G, Eaton R, Hart K, et al. Final report—reflecting environmental land use needs into EU policy: preserving and enhancing the environmental benefits of unfarmed features on EU farmland. Contract No. ENV.B.1/ETU/2007/0033, report for DG Environment. London: IEEP; 2008
5. Luck GW. A review of the relationships between human population density and biodiversity. Biol Rev Camb Philos Soc. 2007;82(4):607–45.
6. Haddaway NR, Bilotta GS. Systematic reviews: separating fact from fiction. Environ Int Elsevier Ltd. 2016;92–93:578–84.

7. Baldock D, Beaufoy G, Bennett G, Clark J. Nature conservation and new directions in the EC Common Agricultural Policy. London: Institute for European Environmental Policy (IEEP); 1993.
8. Meijaard E, Sheil D. The dilemma of green business in tropical forests: how to protect what it cannot identify? Conserv Lett. 2012;5:342–8.
9. Sarkar S, Margules C. Operationalizing biodiversity for conservation planning. J Biosci. 2002;27:299–308.
10. McKinney ML. Urbanization, biodiversity, and conservation. Bioscience. 2002;52(10):883.
11. Morris EK, Caruso T, Buscot F, Fischer M, Hancock C, Maier TS, et al. Choosing and using diversity indices: insights for ecological applications from the German Biodiversity Exploratories. Ecol Evol. 2014;4(18):3514–24.
12. Guerrero I, Morales MB, Oñate JJ, Aavik T, Bengtsson J, Berendse F, et al. Taxonomic and functional diversity of farmland bird communities across Europe: effects of biogeography and agricultural intensification. Biodivers Conserv. 2011;20(14):3663–81.
13. Jetz W, Thomas GH, Joy JB, Redding DW, Hartmann K, Mooers AØ. Global distribution and conservation of evolutionary distinctness in birds. Curr Biol [Internet]. The Authors; 2014;24 (9):919–30. Available from: http://dx.doi.org/10.1016/j.cub.2014.03.011

Chapter 3
Identifying HNV Areas Using Geographic Information Systems and Landscape Metrics

Petra Šímová

Abstract This chapter describes some spatial tools suitable to map and identify High Nature Value (HNV) farmlands, applying geographical information systems (GIS). It provides a list and definitions of some of the most adequate landscape metrics that can be used to identify HNV areas from maps, as well as the main issues that operators can find when handling GIS.

Keywords Landscape metrics • Multispectral images • Edge density • Mean patch size • PCA • FRAGSTATS

3.1 Introduction

Some studies have quantified the correlation between marginal vegetation and the distribution pattern of bird diversity [1, 2] and the relative importance of each type of marginal vegetation for each bird species, confirming how these elements are useful surrogates of High Nature Value (HNV) farmlands [3]. Some species are more related to hedgerows, while others are related primarily to shrubs and others are even related to uncultivated features. Marginal vegetation and isolated patches can be studied easily at the level of landscape spatial composition and configuration by a geographical information system (GIS) approach [4]. These elements promote landscape heterogeneity and can be analysed and measured using several landscape metrics.

Quantification of landscape heterogeneity and other structural characteristics of landscape or habitats using landscape metrics is not a new topic (for example, see O'Neill et al. [5]). However, the research possibilities of metrics application are still evolving due to the ongoing developments in geoinformation technologies and spatial data availability, development of theoretical approaches to measuring of landscape structure and knowledge of possible misuse (see e.g. [6–12].

P. Šímová (✉)
Faculty of Environmental Sciences, Department of Applied Geoinformatics and Spatial Planning, Czech University of Life Sciences Prague, Kamýcká 129, CZ-165 00 Prague 6, Czech Republic
e-mail: simova@fzp.czu.cz

© Springer International Publishing AG 2017   37
F. Morelli, P. Tryjanowski (eds.), *Birds as Useful Indicators of High Nature Value Farmlands*, DOI 10.1007/978-3-319-50284-7_3

In general, landscape structure can be characterized by the composition and spatial configuration of the landscape elements (e.g. [13]). Composition refers to the diversity and frequency of patches within the landscape, irrespective of their spatial configuration. Composition can be expressed by so-called non-spatial metrics, which can be sorted into four basic groups: relative class frequency, richness, evenness and diversity. Landscape configuration, in contrast, is related to spatially explicit characteristics of land cover types. The principal aspects of a landscape's spatial configuration are the size distribution and density, shape complexity, core area, isolation, contrast, dispersion, contagion, subdivision and connectivity of the landscape patches. Both composition and configuration metrics are usable not only in landscape studies but also in animal ecology, for example as explanatory variables in species distribution modelling [14–16].

Description of landscape structure, as well as habitat structure, can be performed at three hierarchical levels [13]: for each patch in the landscape mosaic (patch level); for each patch class in the landscape (class level); or for the landscape mosaic as a whole (landscape level). Some metrics are meaningful only at one level (e.g. landscape heterogeneity indices such as the Shannon diversity index (SHDI), Shannon equitability index (SHEI) and Simpson diversity index (SIDI) can be determined only at the landscape level because they are integrated over all patches and classes in the study area), but most are defined for multiple levels (e.g. patch density (PD) and edge density (ED)). For many research topics, including identification of HNV areas, combination of these levels can bring good results, even if only the simplest composition metrics are used (see Fig. 3.1 and the *landscape metrics case study* at the end of this chapter).

The critical issue of calculating and applying landscape metrics is working with an appropriate scale of data and analysis. Most of the landscape metrics are scale dependent, and many authors have pointed out that the scale of the data (observations) and the scale of the analysis must be coherent in order to calculate and interpret these metrics correctly [11, 13, 14, 17–19]. The issue of scaling is related to the grain size (pixel size) or to the minimum mapping unit (MMU), the extent of the study area and the thematic resolution (see Fig. 3.3 for an example). The behaviours of landscape metrics in relation to scaling factors has implications especially for the fields of research and the operations within which these metrics are used as explanatory variables or indicators, for example for species distribution modelling [14–16]. As HNV farmlands are closely connected with the occurrence of hedgerows, tree lines, patches of tall herb vegetation, grassy belts, etc. (i.e. elements that usually create small near-natural patches or green veining within the agricultural matrix), the issue of scale is extremely important in studies focused on HNV landscape indicators. Using an inappropriate scale can lead to inappropriate research results and conclusions (see Šímová and Gdulová [11] for a review of scale effects on metrics behaviour).

Inasmuch as all types of metrics are computed using GIS approaches and a lot of data sources and software can be used, another important issue for obtaining appropriate results is knowledge of the input data, metrics formulas and the basics of the software functionality. Even if packages specializing in landscape metrics are

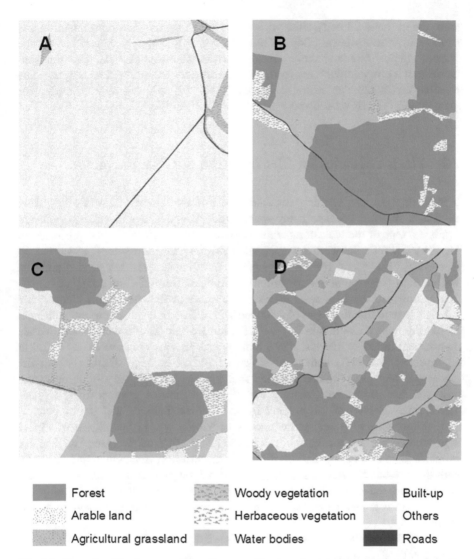

Fig. 3.1 Example of landscapes with various combination of patch richness–Shannon diversity index–edge density (*PR–SHDI–ED*) values. (**a**), (**b**): Landscapes with similar ED and PR, but very different SHDI. (**a**) PR = 5, SHDI = 0.182, ED = 75 m/ha. (**b**) PR = 6, SHDI = 0.900, ED = 80 m/ha. (**c**), (**d**): Landscapes with similar SHDI and PR, but different ED. (**c**) PR = 8, SHDI = 1.083, ED = 121 m/ha. (**d**) PR = 9, SHDI = 1.176, ED = 263 m/ha. (**d**) Heterogeneity is increased not only by near-natural elements like *forests* and *woody* and *herbaceous vegetation*, but also by paved roads and built-up areas. Data and scale: Visual interpretation and manual vectorization of orthorectified aerial photographs (orthophotos), sample localities 1 × 1 km in the Czech Republic. *Author: Petra Šímová*

used, the user should be familiar with the basics of spatial data representation and methods of spatial analysis. Some pitfalls for computing and interpreting landscape metrics in HNV farmland studies, resulting from the issues of scale, data representation and computational methods, are introduced below in this chapter. The possible errors are demonstrated especially for the simplest metrics usable for identification of HNV farmland.

3.2 Simple Metrics for Identification of HNV Farmland

It is not the purpose of this chapter to give a list and description of all existing landscape metrics, because many specialized extensive works on this topic are available. One of the most comprehensive overviews of the metrics is presented in the FRAGSTATS software documentation [13], where the reader is given a deep theoretical background for using metrics in landscapes and ecological studies, as well as their description, interpretation, limitations, formulas and domains.

As shown in the *landscape metrics case study* at the end of this chapter, the simplest way to identify types of farmland with different levels of natural value can be to employ a combination of areas of main habitat (land cover) types, density metrics and diversity metrics. This simple approach does not have to be fruitful in all types of landscape and at all scales or for all ecological processes in the landscape, and use of more complex metrics can be necessary in some cases; however, use of the area, density and diversity metrics appears to be logical. These categories cover the amount of near-natural habitats and the amount of agricultural matrix, as well as the variety of habitat types (diversity metrics) and their arrangement in smaller or bigger elements (density metrics) (see Fig. 3.1c, d). In addition, a strong correlation of species occurrence, abundance and diversity with the amount of preferred habitat, size of habitat patches and area of ecotones has been proved in hundreds of studies. An undeniable advantage of using the simple metrics is also their simple interpretation and predictable behaviour across scales [11].

Description of the metrics used in the landscape metrics case study (based on McGarigal [13]; adjusted).

Percentage of Landscape refers to how much of the landscape is composed of a particular habitat type. The percentages of *forests, non-forest woody vegetation* (e.g. tree lines, hedgerows, groups of trees and shrubs), *non-agricultural herbaceous vegetation* (e.g. near-natural meadows, herbaceous edges of forests, roads, streams and waterbodies), *urban areas* (small villages in this case), *agricultural grasslands* and *arable land* were used.

Mean Patch Size (MPS) is a function of the number of patches and the total area, which can be calculated at the landscape level and the class level. In the sense of GIS terminology, the total number of polygons is divided by the total area of the locality (at the landscape level), or the number of polygons of a certain class is divided by the sum of the polygon areas of the same class (at the class level). MPS at the landscape level was calculated in our *landscape metrics case study*; however, MPS at the class level could also be useful in the HNV context.

Edge Density (ED) is also meaningful at both the class and landscape levels. It is calculated as the sum of the lengths of the edges of the patches (polygons) of a certain habitat class or as the sum of all polygon edges. Depending on the type of landscape and landscape elements contained in the input data, ED can express a measure of the size of the patches and the area of ecotones in the landscape, as well as the number of linear elements (Fig. 3.1). One of the known limitations is that the ED value is affected by the resolution of the input data. Generally, the greater the detail with which edges are delineated in the data, the greater the edge length.

Patch Density (PD) is the inverse value of MPS, calculated as the total area of the locality divided by the number of polygons (at the landscape level) or the sum of polygon areas of the same class divided by the number of polygons of that class (at the class level). PD and MPS are usually highly (positively and negatively, respectively) correlated with ED, but not necessarily. In general, landscape elements of the same size can have greatly varying edge lengths. Hence, the correlation depends on the shapes of the landscape elements and the jaggedness of their edges.

Patch Richness (PR) is the simplest measure of landscape heterogeneity, which states the number of classes (habitat types) present within the study area. Like all diversity metrics, it is computed only at the landscape level. PR is not affected by the area of each class; therefore, two landscapes of the same PR value may have very different structures (see Fig. 3.1).

Shannon diversity (SHDI) is one of the most popular diversity indices. In origin, it was introduced by information theory [20] and further applied in ecology of communities [21]; however, it is also a frequently used measure of heterogeneity in landscape ecology. Shannon diversity is computed as

$$SHDI = -\sum_{i=1}^{n} (p_i * \ln p_i)$$

where n is the number of classes (habitat types) present in the study area (i.e. PR), p_i is the proportion of the study area occupied by each class (habitat type) i. Proportions p_i are within the interval 0.1, hence the natural logarithm of p_i is negative. Therefore, the whole expression is multiplied by (-1) to obtain a positive value, thus a higher heterogeneity is expressed by a higher SHDI. As follows from the formula, the value of the SHDI increases not only with the number of classes (PR), but also with the evenness of their proportions. From two landscapes with the same PR, the higher SHDI will apply to the landscape where the proportions of classes are more similar. For example, the SHDI of a landscape composed of five habitat types with the proportions 0.1, 0.1, 0.1, 0.1 and 0.5 equals 1.27, while a landscape with the same habitat types arranged in the proportions 0.2, 0.2, 0.2, 0.2, 0.2 reaches an SHDI equal to 1.61.

Simpson diversity (SIDI) is another popular measure of heterogeneity borrowed from community ecology. Similarly to Shannon diversity, SIDI works with proportions of classes for the purposes of landscape ecology and as a result it combines richness and evenness of habitat classes. Its interpretation is quite intuitive: the SIDI value given the formula below represents the probability that any two random points would fall in different habitat types.

$$SIDI = 1 - \sum_{i=1}^{n} p_i^2$$

All three diversity metrics have several important limitations in terms of interpretation and the comparability of their values across landscapes. One of the usual criticisms is that the measures do not convey any information about habitat quality. Due to this fact, landscape heterogeneity can increase not only owing to the phenomenon of a rising proportion of near-natural areas and green veining in a uniform agricultural landscape, but, for example, also due to spatial development of urbanized or mining areas, to the detriment of near-natural habitats (see Figs. 3.1 and 3.2). Hence, HNV farmland may theoretically have lower values of diversity indices than farmland without any ecological value. From the scaling point of view, the metrics values may depend on the study area extent, and there can be more habitat types in larger areas of interest. Moreover, spatial data describing the same study area can be created in different ways and for various purposes, so the same locality can be represented by completely different data sets, varying not only in spatial resolution but also in thematic resolution (see Fig. 3.3). This means that close attention should be paid to the character and extent of the area under study and to the way in which the landscape is represented in the input spatial data.

3.3   Input Data for Calculation of Landscape Metrics

Generally, two types of input data can be taken as basic information sources for computing landscape metrics: data acquired by remote sensing technologies (aerial or satellite images) and various types of maps obtained by ground mapping. In any case, categorical raster or vector layers of land cover (habitat) types must be created prior to beginning GIS analyses or using specialized software such as FRAGSTATS, Patch Analyst or V-Late.

As landscape metrics are often scale dependent and important elements of HNV farmlands are often composed from small patches or linear elements, input data at a detailed scale should be used as basic input for HNV analysis. As Fig. 3.4 shows, commonly available multispectral satellite data such as Landsat and Sentinel images have too coarse a resolution (30 m and 10 m, respectively) for recognizing small-sized landscape elements; therefore, they can be useful only for work at scales of tens or hundreds of square kilometres, where small patches and lines are disregarded. Input data with spatial resolution fine enough to describe landscapes including small groups of bushes or trees, near-natural herbaceous patches, etc., or

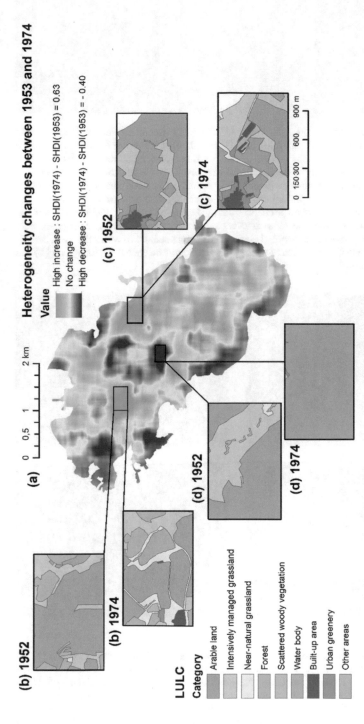

Fig. 3.2 Example of different reasons for changing Shannon diversity index (*SHDI*) values in farmland: changes in spatial heterogeneity in the Kremze basin (Czech Republic) between 1952 and 1974. (**a**) Map of heterogeneity changes. The raster values were computed by subtracting two grids of spatial heterogeneity expressed by SHDI: SHDI (1974) minus SHDI (1952). (**b**) Example of greatly increasing heterogeneity. The area was diversified due to building of a small fishpond with near-natural grassland at its edge and division of arable land by non-cultivated grassy belts. (**c**) However, heterogeneity expressed in this way can increase greatly also due to development of urban areas and to changing spatial patterns of agricultural areas. (**d**) Example of greatly decreasing landscape heterogeneity. Although in 1952 the plot was quite intensively managed, liquidation of scattered vegetation and transformation of surrounding grasslands into arable land caused a great decrease in heterogeneity. *Authors: Petra Šímová, Eva Bártová*

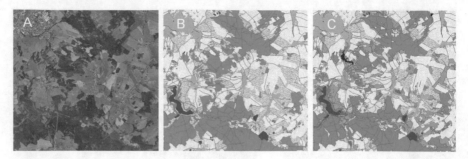

Fig. 3.3 Example of input data preparation and influence of thematic resolution. The input data for calculation of landscape metrics are categorical layers (**b, c**) of land cover (habitats). One of the methods to create categorical maps is visual interpretation and manual vectorization of aerial photographs (**a**) (combined with field mapping in this case). As the method is subjective, it is necessary to prepare an interpretation methodology in advance. The main part of the methodology should be an interpretation key, describing all the categories that will be distinguished. The thematic detail (thematic resolution) influences many landscape metrics. Pictures (**b**) and (**c**) document two vectorizations of the same landscape: (**b**) only the main land cover types were distinguished; (**c**) more types of near-natural elements were mapped. It is obvious that area, density and diversity metrics would give different results and could indicate another landscape for (**b**) and (**c**). *Author: Petra Šímová*

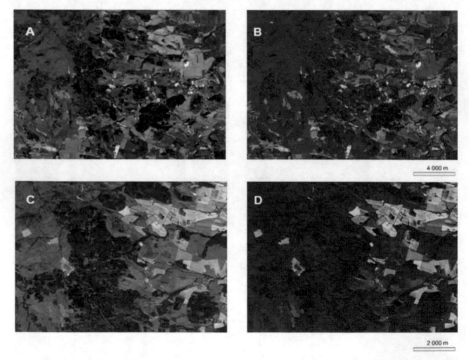

Fig. 3.4 Freely available multispectral satellite images: *Landsat 8* in natural (**a**) and false (**b**) colours, resolution 30 m. At this resolution, small-sized elements in the landscape are not depicted. Better resolution (10 m) for analysis of High Nature Value (*HNV*) farmlands is given by *Sentinel 2A* images (**c, d**). Methods of classification of multispectral images enable us to reach more detailed thematic resolution (i.e. obtain categorical maps distinguishing more habitat types) in comparison with visual interpretation methods. *Authors: Tomáš Klouček, Jiří Prošek, Petra Šímová*

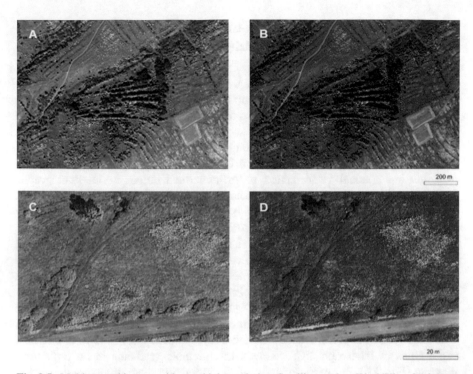

Fig. 3.5 Multispectral images with very high resolution. Satellites such as WorldView 2 (pictures (**a**), (**b**), image in natural and false colours, respectively) with resolution of 2 m enables us to recognize small woody elements and various types of herbaceous habitats in the landscape under study. For more detailed resolution (10 cm in this case) it is possible to use unmanned aerial vehicles (drones) equipped with a multispectral camera. Picture (**c**) was obtained using a hexacopter with a Tetracam camera. Individual bushes and trees can be recognized, as well as a lot of types of herbaceous vegetation (better distinguishable on the false colour version (**d**)). The advantage of such data is combination of high spatial and spectral resolution. The disadvantage is the cost of the images. *Authors: Tomáš Klouček, Jiří Prošek, Jan Komárek, Ondřej Lagner, Petra Šímová*

even individual trees or bushes, can be acquired by satellites with very high resolution (VHR), e.g. WorldView 2 and 3, or QuickBird, or by using unmanned aerial vehicles (UAVs, drones) equipped with a multispectral camera (see Fig. 3.5).

An important advantage of utilizing remote sensing multispectral data is that various classification methods (e.g. supervised or non-supervised classification [22]) can be used to extract the categorical land cover information. Such classifications are not as dependent on subjective perception by a person or operator as manual methods of data preparation. Moreover, more detailed thematic information can be obtained (more habitat types can be recognized) thanks to spectral resolution of the images and advanced classification methods than by using visual interpretation only. If panchromatic or colour (RGB) images (usually aerial photographs) are taken as an input, the spectral information is too limited to separate land cover classes by automatic methods [23]. Therefore, visual interpretation and time-consuming manual vectorization of aerial photos are usually used in practice [23–25], sometimes in combination with field surveys or with existing maps.

3.4 Landscape Metrics Case Study: An Example of Using Landscape Metrics to Identify HNV Farmland

This short case study shows an example of how landscape metrics (even the simplest) can help to identify HNV farmlands at a detailed scale of analysis. The study used 358 sample localities (squares of 1×1 km) randomly distributed within farmland in the Czech Republic. The samples varied from completely uniform arable land to highly diversified (mountainous) mosaics of meadows, pastures and patches of forests. The input categorical layers were prepared using visual inter-pretation and manual vectorization of orthorectified colour aerial photographs (orthophotos) according to a detailed vectorization methodology and interpretation key. The categories distinguished were the percentages of *forests*, *non-forest woody vegetation* (e.g. tree lines, hedgerows, groups of trees and shrubs), *non-agricultural herbaceous vegetation* (e.g. near-natural meadows, herbaceous edges of forests, roads, streams and waterbodies), *urban areas* (small villages in this case), *water bodies*, *agricultural grasslands* and *arable land*. Simple landscape metrics (the percentages of the categories, ED and MPS at the landscape level, and the diversity metrics PR, Shannon diversity and SIDI) were computed using the basic tools of ArcGIS and MS Excel. Principal component analysis (PCA) in Canoco 5 software was employed to explore whether the landscape metrics were able to distinguish some types of landscape. PCA showed interpretable gradients in an ordination diagram (Fig. 3.6). For each quadrant of the ordination diagram, examples of the best-fitting sample landscape are presented (Figs. 3.7, 3.8, 3.9 and 3.10).

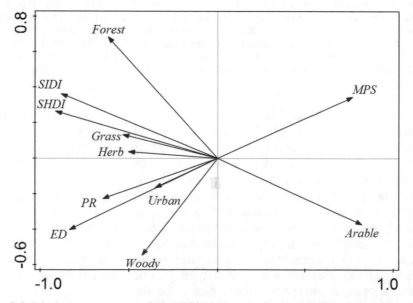

Fig. 3.6 Principal component analysis (*PCA*) with area, density and diversity metrics. Area metrics: percentages of forests (*Forest*), non-forest woody vegetation (*Woody*), non-agricultural herbaceous vegetation (*Herb*), urban areas (*Urban*), agricultural grasslands (*Grass*) and arable land (*Arable*). Density metrics: mean patch size (*MPS*) and edge density (*ED*). Diversity metrics: patch richness (*PR*), Shannon diversity index (*SHDI*) and Simpson diversity index (*SIDI*)

Fig. 3.7 Typical landscapes of the first quadrant (*Forest*, Shannon diversity index (*SHDI*), Simpson diversity index (*SIDI*)). The typical samples are characterized by a mosaic of forests, non-forest woody patches, agricultural grasslands and patches of herbaceous vegetation. The metrics identified High Nature Value (*HNV*) farmland with a patchiness pattern

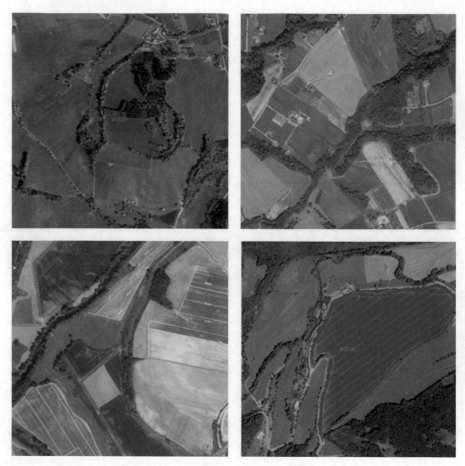

Fig. 3.8 Typical landscapes of the third quadrant (patch richness (*PR*), edge density (*ED*), *Woody*). The typical samples are characterized especially by occurrence of linear elements of woody vegetation in agricultural matrix, with addition of small-sized woody and herbaceous patches. Hence, the metrics identified High Nature Value (*HNV*) farmland with a significant proportion of linear elements

Fig. 3.9 Typical landscapes of the second quadrant (mean patch size (*MPS*)). The typical samples are characterized especially by big patches of arable land combined with patches of forests or woody vegetation. MPS (strongly negatively correlated with edge density (*ED*)) helps to interpret samples of the second quadrant as a quite uniform coarse grain (in relation to the scale of analysis) agriculture landscape with occurrence of forest patches

Fig. 3.10 Typical landscapes of the fourth quadrant (*Arable*). Agricultural landscapes with an overwhelming majority of arable land. The uniformity of the sites is sometimes interrupted by rare scattered woods or linear elements

References

1. Ceresa F, Bogliani G, Pedrini P, Brambilla M. The importance of key marginal habitat features for birds in farmland: an assessment of habitat preferences of red-backed shrikes *Lanius collurio* in the Italian Alps. Bird Study [Internet]. BTO; 2012 Aug 13 [cited 2015 Feb 15];59(3):327–34. Available from: http://www.tandfonline.com/doi/abs/10.1080/00063657. 2012.676623#.VOSmWPmG_X4
2. Regos A, Domínguez J, Gil-Tena A, Brotons L, Ninyerola M, Pons X. Rural abandoned landscapes and bird assemblages: winners and losers in the rewilding of a marginal mountain area (NW Spain). Reg Environ Chang. 2016;16(1):199–211.
3. Morelli F. Relative importance of marginal vegetation (shrubs, hedgerows, isolated trees) surrogate of HNV farmland for bird species distribution in Central Italy. Ecol Eng [Internet].

Elsevier B.V.; 2013 Aug [cited 2013 May 16];57:261–6. Available from: http://linkinghub. elsevier.com/retrieve/pii/S0925857413001614

4. Bennett AF, Radford JQ, Haslem A. Properties of land mosaics: implications for nature conservation in agricultural environments. Biol Conserv [Internet]. 2006 Nov [cited 2012 Oct 4];133 (2):250–64. Available from: http://linkinghub.elsevier.com/retrieve/pii/S0006320706002850

5. O'Neill RV, Krummel JR, Gardner RH, Sugihara G, Jackson B, DeAngelis DL, et al. Indices of landscape pattern. Landsc Ecol. 1988;1(3):153–62.

6. Wu J, Shen W, Sun W, Tueller PT. Empirical patterns of the effects of changing scale on landscape metrics. Landsc Ecol. 2002;17(8):761–82.

7. Shao G, Wu J. On the accuracy of landscape pattern analysis using remote sensing data. Landsc Ecol. 2008;23(5):505–11.

8. Riitters K, Vogt P, Soille P, Estreguil C. Landscape patterns from mathematical morphology on maps with contagion. Landsc Ecol. 2009;24(5):699–709.

9. Kelly M, Tuxen KA, Stralberg D. Mapping changes to vegetation pattern in a restoring wetland: finding pattern metrics that are consistent across spatial scale and time. Ecol Indic. Elsevier Ltd; 2011;11(2):263–73.

10. Saura S, Vogt P, Velázquez J, Hernando A, Tejera R. Key structural forest connectors can be identified by combining landscape spatial pattern and network analyses. For Ecol Manag. 2011;262(2):150–60.

11. Šímová P, Gdulová K. Landscape indices behavior: a review of scale effects. Appl Geogr [Internet]. 2012 May [cited 2014 Sep 2];34(3):385–94. Available from: http://linkinghub. elsevier.com/retrieve/pii/S0143622812000057

12. Lausch A, Blaschke T, Haase D, Herzog F, Syrbe RU, Tischendorf L, et al. Understanding and quantifying landscape structure—a review on relevant process characteristics, data models and landscape metrics. Ecol Modell. Elsevier B.V.; 2015;295:31–41.

13. McGarigal K. Fragstats.Help.4.2. 2015. p. 1–182.

14. Bailey D, Herzog F, Augenstein I, Aviron S, Billeter R, Szerencsits E, et al. Thematic resolution matters: indicators of landscape pattern for European agro-ecosystems. Ecol Indic. 2007;7(3):692–709.

15. Benedek Z, Nagy A, Rácz IA, Jordán F, Varga Z. Landscape metrics as indicators: quantifying habitat network changes of a bush-cricket *Pholidoptera transsylvanica* in Hungary. Ecol Indic. 2011;11:930–3.

16. Gao J, Li S. Detecting spatially non-stationary and scale-dependent relationships between urban landscape fragmentation and related factors using geographically weighted regression. Appl Geogr Elsevier Ltd. 2011;31(1):292–302.

17. Li H, Wu J. Use and misuse of landscape indices. Landsc Ecol. 2004;19(4):389–99.

18. Wu J. A landscape approach for sustainability science. In: MP W, RE T, editors. Sustainability science: the emerging paradigm and the urban environment. New York: Springer; 2012. p. 441.

19. Wheatley M. Domains of scale in forest-landscape metrics: implications for species-habitat modeling. Acta Oecologica Elsevier Masson SAS. 2010;36(2):259–67.

20. Shannon CE. The mathematical theory of communication. Bell Syst Tech J. 1948;27:379–423.

21. Pielou EC. Ecological diversity. New York: Wiley-Interscience; 1975.

22. Lillesand T, Kiefer RW, Chipman J. Remote sensing and image interpretation. 7th ed: Wiley; 2015.

23. Sluiter R, Jong SM. Spatial patterns of Mediterranean land abandonment and related land cover transitions. Landsc Ecol. 2007;22(4):559–76.

24. Baessler C, Klotz S. Effects of changes in agricultural land-use on landscape structure and arable weed vegetation over the last 50 years. Agric Ecosyst Environ. 2006;115(1–4):43–50.

25. Sklenicka P, Šímová P, Hrdinová K, Salek M. Changing rural landscapes along the border of Austria and the Czech Republic between 1952 and 2009: roles of political, socioeconomic and environmental factors. Appl Geogr. 2014;47:89–98.

Chapter 4
Suitable Methods for Monitoring HNV Farmland Using Bird Species

Piotr Tryjanowski and Federico Morelli

Abstract In this chapter we summarize a list of technical and statistical approaches, useful for studying High Nature Value (HNV) farmlands, and biodiversity components. We also focus on the use of suitable frameworks to monitor HNV farmland using bird species, providing some information about bird data collection in the field. The concept of bioindicators and the use of bird species as environmental surrogates are briefly described, as well as the most common diversity metrics used to describe bird communities (taxonomic diversity, functional diversity and phylogenetic diversity). Finally, we provide some suggestions about statistical methods that can be followed in order to link the study of HNV farmlands to biodiversity patterns and environmental characteristics of farmlands.

Keywords Bird data collection • Point count • Bioindicator • Species distribution model SDM • MRT • IndVal analysis

4.1 Bird Count Methods for Farmland Systems: Single-Habitat and Multi-habitat Species

Piotr Tryjanowski

To establish the importance of High Nature Value (HNV) farmland for bird populations, qualitative data on particular bird species' presence, as well as quantitative information on local bird population sizes, are crucial. However, the possibility to collect sufficient data is related to human resources, time, money

P. Tryjanowski (✉)
Institute of Zoology, Poznań University of Life Sciences, Wojska Polskiego 71 C, PL-60-625 Poznań, Poland
e-mail: piotr.tryjanowski@gmail.com

F. Morelli (✉)
Faculty of Environmental Sciences, Department of Applied Geoinformatics and Spatial Planning, Czech University of Life Sciences Prague, Kamýcká 129, CZ-165 00 Prague 6, Czech Republic
e-mail: fmorellius@gmail.com

© Springer International Publishing AG 2017   53
F. Morelli, P. Tryjanowski (eds.), *Birds as Useful Indicators of High Nature Value Farmlands*, DOI 10.1007/978-3-319-50284-7_4

and the purposes of research programmes [1]. For the whole-continent scale it is very popular to use data from the Pan-European Common Bird Monitoring Scheme (PECBMS) organized by the European Bird Census Council (EBCC) and BirdLife International (http://www.ebcc.info). This scheme collates population trend data from annually operated national breeding bird surveys from across Europe. The focus is on population trends of widespread birds and the scheme aims to promote the use of birds as 'bioindicators' of the state of nature and of the health of the environment. However, these data are not dedicated to particular HNV farmland areas, because they were established for general monitoring purposes and the researchers chose between many different variants of bird count methods.

The most commonly used methods can be divided into three categories—from the most to the least time consuming: (1) mapping, (2) transects and (3) point counts. However, for statistical purposes, links to environmental data are not necessary to use the first one, and in the case of transects and points it is possible to cover broader ranges of the areas and to obtain more representative sample sizes.

In European conditions, mapping of birds starts in April and finishes in July; however, it can differ in local conditions due to changes in the local birds' activity patterns. The chosen areas (plots, sometimes regular squares) are surveyed 6–10 times, mainly during mornings, but additionally with visits at night to detect nocturnal active species (like *Crex crex* (see more details below) or *Acrocephalus*, *Locusella* and *Luscinia* species). The detected birds are plotted on maps and after the whole breeding season, territories are marked. Because the method is very time consuming, it is not very popular in current monitoring schemes and in establishing HNV farmland importance for birds. However, using it with modifications (a more simple design, especially a limited number of visits per study plot) is very valuable especially in more complex analyses [2, 3].

A much easier and less time-consuming method is data collection by the transect method. Sometimes transects, especially for monitoring purposes, are located in 1 km^2 grid cells [4]. During the breeding season each plot is surveyed twice. In central European conditions the first visit takes place between 10 April and 15 May and the second between 16 May and 30 June. The bird census within each square consists of two parallel 1 km transects along either an east–west or a north–south axis. Each transect is divided into five 200 m sections, in which birds are noted within three distance categories (100 m). Birds are noted perpendicular to the transect line. Each survey starts between the dawn and 9 am and lasts for about 90 min. The surveys are carried out by volunteers.

In recent times, even in studies presented in further chapters of this book, more often data collection has been based on results from point counts. For evaluation of conservation tools in Poland, a large data set was used by Żmihorski et al. [5], where detailed information on the point count method was also provided. It is worth noting because it fits very well with the general methodology commonly used also to establish the influence of HNV farmland on bird populations. The point count is a relatively simple method commonly used also for nationwide monitoring programmes. All sites chosen for bird counts are visited twice in the breeding season, once in early spring (April–May) and once in late spring (the end of May–

Fig. 4.1 Schematic representation of the bird point count method, used to record and identify bird species at a sample site. The fixed radius is defined a priori and usually is 100 m around the observer (*yellow dashed line*). The location of each point is recorded with a GPS. On the field, the trained observer has to record all individuals of all species of birds seen or heard around the survey point

June) during early mornings (from sunrise to 10:00 a.m. during the first visit and to 9:00 a.m. during the second visit). Data are collected only on mornings without strong wind or rain. Each visit lasts 10 min (in some schemes 5 min), and all seen and heard individuals of bird species within a 100 m radius are counted, except for nestlings and birds passing by more than 50 m above the ground (these are considered as not breeding at the site) (Fig. 4.1). The same is true at non-breeding times, when above-passing birds are not directly connected to the study plot (e.g. high-flying migrants).

Although the different methods generate slightly different results, fortunately in quite simple farmland habitats the obtained findings—mainly basic measures like the number of species, diversity indices and general information on the density— are strictly correlated [6]. But sometimes it is necessary to use a more advanced method dedicated to particular species, which is crucial for secretive birds, mainly with nocturnal activity. One spectacular example, especially in the context of HNV importance for declining farmland specialists, is the corncrake, *Crex crex*. Fortunately Budka and Kokociński [7] examined the accuracy of three counting methods—territory mapping, point-based censusing and point counting—in evaluating the population size and spatial distribution of the corncrake.

They found significant differences among the three methods in the number and distribution of males counted within the study plots. The middle point count approach consistently underestimated the population size regardless of the number of males present, while the point-based census approach overestimated the population size when a small number of males were present within the study plot, but underestimated it when males were numerous. However, the conclusions directionally say that the accuracy of the counting methods may depend on bird density. Moreover, point-based censusing may prove quite inaccurate in pinpointing the locations of calling males. When censusing species with long-range acoustic communication, like the corncrake, it is challenging to estimate the distance to calling birds, which means that the standard point count method should be applied carefully.

All methods mentioned above, albeit with caution, can be used to underline HNV farmland importance to birds. However, it is worth noting that classical monitoring schemes, as well as others based on volunteer help, are not dedicated, and hence not perfectly suited, to HNV farmland.

4.2 The Concept of Bioindicators or Environmental Surrogates and Common Measures of Diversity in Bird Communities

Federico Morelli

4.2.1 The Concept of Bioindicators

The use of surrogates or bioindicators is like the use of shortcuts in ecology: a cost-effective strategy in order to study very complex systems [8, 9]. Bioindicators are potential indicators of ecological integrity at various levels of the ecosystem. Following the review of Carignan and Villard [10], the use of bioindicators is particularly indicated when (a) many species representing various taxa and life histories are included in the monitoring programme; [2] their selection is primarily based on a robust quantitative database from the focal region; and [3] caution is applied when interpreting their surrogate power to mirror the deterioration of ecological integrity.

Surrogates can be roughly divided into two categories: taxonomic (biotic) and environmental (abiotic) surrogates. Taxonomic surrogates are predominantly based on biological data—for example, groups of species such as birds [11]. Among the numerous 'surrogate' candidates developed in recent decades, bird distribution is potentially one of the most useful because birds are widely distributed, and breeding bird records are among the easiest species distribution data sets to obtain, thanks to the presence of birding all around the world [12, 13].

For a more detailed and complete explanation about types of bioindicator, it is possible to read Caro [14].

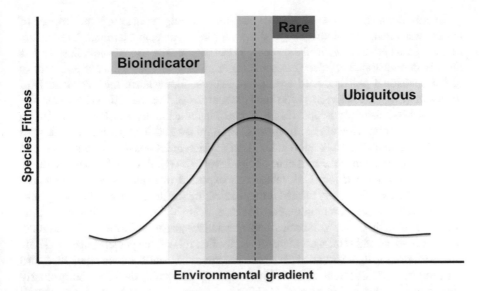

Fig. 4.2 Three types of species classified as ubiquitous species, rare species and a (potential) bioindicator, on the basis of their environmental tolerance. The scheme is based on Holt and Miller [19]

As a summary, many concepts can be included as biological or environmental surrogates: from flagship species to umbrella species, focal species, indicator species, sentinel species, biomonitoring species, biomarker species and many others [8, 15–17]. The use of bioindicators or proxies to assess ecosystem responses and key ecological process changes is a central point in ecology because of resource constraints in monitoring and management, requiring cheap and effective ways [8, 13, 18]. However, the selection of bioindicators has to pay attention to some basic principles, related to the sensitivity of the species focused to a gradient of environmental changes. Rare species and too generalist or ubiquitous species are not suitable candidates as bioindicators. In this regard, Fig. 4.2 is an adaptation from Holt and Miller [19].

Then, a good bioindicator should be sufficiently sensitive to provide an early warning of environmental change, widely applicable, able to provide a continuous assessment over a wide range of stress factors, relatively independent from the size of the sample, easy and cost effective to measure or collect, able to discriminate between natural cycles or trends and those that are anthropogenic stress induced, and relevant to ecologically significant phenomena [17, 20, 21].

Farmland birds have been among the species considered most at risk for conservation in recent decades, considering that farmland landscapes represent more or less 45% of the total European land area (Food and Agriculture Organization; http://apps.fao.org/). Bird species have been used as bioindicators in many studies [12, 22–24] because the presence of some species is strongly correlated to different characteristics of environments [24–29].

In this context, the use of bird indicators can help to quantify the impact of agricultural change on biodiversity and monitor the change, in a temporal dimension, of the characteristics of farmland areas. Furthermore, the use of bioindicators has already demonstrated efficiency in preventing biodiversity loss, mainly as a tool to speed decisional policies. Bird species indicators, for instance the Farmland Bird Indicator (FBI), are widely used in many countries of Europe [30, 31] because the FBI has been formally adopted as a structural indicator by the European Union (EU 2005; Commission of the European Communities 2006). Other useful indicators are the community specialization index (CSI), which measures the average degree of habitat specialization at the bird community level [32–34]. A study by Devictor et al. [35] [32] showed that low CSI values are expected in disturbed and fragmented habitats, such as intensively farmed areas, indicating the dominance of habitat generalists within the local community. For instance, in HNV farmland one can expect the presence of both generalist species and specialist species because HNV farmland provides natural and semi-natural vegetation, often related to specialist species [32].

Furthermore, in a recent study from Germany, scientists suggested that bird communities are different in each kind of HNV farmland, distinguishing clearly among at least three main groups: (1) HNV with prevalence of hedgerows or small forest patches; (2) wet grasslands; and (3) open agricultural land of low land use intensity [36]. Then, different bird species can characterize different HNV farming systems.

These indications are useful in order to identify the bird species that are better candidates as surrogates of HNV farmland integrity, or potentially HNV farmland bioindicators. For instance, in some studies, only species with an occurrence equal to or higher than a given threshold are proposed as bioindicators, also in order to guarantee statistical robustness [25, 37]. Furthermore, the habitat or diet specialization of birds inhabiting the farmland or grasslands is important to consider during the bird species selection procedure [38].

4.2.2   Common Diversity Metrics Useful for Bird Communities

Different diversity metrics can be used as surrogates of biodiversity in bird communities. Here we present a short list, with a very brief explanation. For a more complete treatment we suggest reading Laureto et al., Luck et al., Guilhaumon et al., Santini et al. and Tucker et al. [39–43]

- **Taxonomic diversity (TD)** [44] is expressed as bird species richness or the number of recorded bird species in each community:

$$S = N_{tot} = no.species$$

where N_{tot} is the number of species found in a community.

- **Community specialization index (CSI)** is an index that measures the average degree of habitat specialization among the individuals of the community (on the basis of the occurrence of each one in different types of available environments) as the arithmetic mean of the species specialization index (SSI) weighted by the abundances [33, 35]:

$$CSI = \sum_{i=1}^{n} \frac{N_i}{N_{\text{tot}}} * SSI_i$$

where SSI_i is the specialization index for each species, $_i$, weighted by its abundance, N_i, and divided by the summed abundances of all species, N_{tot}.

- **Community trophic index (CTI)** is an index that discriminates, for example, between communities with more granivorous species and communities with more insectivorous and carnivorous species, because CTI is determined on the basis of the diet proportions [31].
- **Functional diversity (FD)** is calculated using avian niche traits related to feeding and breeding ecology [45]. The biodiversity metrics based on species–trait approaches are focused on functional aspects of biodiversity and constitute an additional tool to the traditional taxonomic approach [46]. Different FD indices can be estimated for each community. Petchey's FD is calculated with bird traits, using 'vegan', 'ade4' and 'picante' packages. Petchey's FD measures diversity by constructing a dendrogram representing the similarity/dissimilarity among species of the entire community, based on the available traits, and then calculates the total branch length of individual communities superimposed on the dendrogram [47, 48]. Another possibility is represented by the use of multidimensional FD indices [49]. Three independent FD measures that each capture one of the three primary components of FD are used to describe the overall FD in an assemblage: functional richness (FRic) represents the amount of functional space occupied by a species assemblage; functional evenness (FEve) indicates the regularity of the degree to which the biomass of the species assemblage is distributed in a niche space to allow effective utilization of the entire range of resources available; and functional divergence (FDiv) defines how far high species abundances are from the centre of the functional space.
- **Community evolutionary distinctiveness (CED)** is a measure of species uniqueness, measured as the 'evolutionary distinctiveness' (ED) score [50], using a phylogeny or evolutionary tree. The phylogenetic diversity is estimated using the sum of the branch length of the species present in the assemblage [51]. The ED scores for each species are calculated dividing the total phylogenetic diversity of a clade among its members [52]. The CED or average ED in a given community or assemblage is calculated as:

$$CED = \sum_{i=1}^{n} \frac{ED_i}{N_{\text{tot}}}$$

where ED_i is the evolutionary distinctiveness score for each species, i, divided by the number of all species recorded in the community, N_{tot}.

- **Phylogenetic diversity (PD)** is estimated on the basis of a tree that represents the phylogenetic relationships between species. The species that are closely related and are very similar with respect to phylogenetics are placed next to each other [53].

4.3   Species Distribution Models and Other Useful Statistical Tools

Federico Morelli

Models are increasingly being used as key components of wildlife management programmes, because they provide a method to predict the outcomes of management and conservation strategies. Ecological models traditionally privilege generality rather than accuracy, looking for simplicity [54]. Niche modelling, also known as species or habitat potential distribution modelling [55, 56], is used to inductively interpolate or extrapolate fundamental niches outside the locations where a species is present, by relating species presence to environmental predictors [57]. The large majority of the available wildlife data sets consist of presence-only data sets, coming from atlases, museum and herbarium records, observational databases and in situ field surveys [58–60].

Species distribution models (SDMs) are numerical tools that combine observations of species occurrence or abundance with environmental estimates [61, 62]. Predictive SDMs [63] are increasingly applied as extrapolative tools for purposes of conservation planning and management of ecosystems. Such models rely on the concept of the ecological niche occupied by the detected species. But SDMs are more than just simple predictive models of species' suitable habitats—they are an important tool for conservation [55, 64].

4.3.1   SDMs in a Nutshell

The procedures related to SDMs are extensively explained in Franklin [65].
   A model procedure can be summarized briefly as follows:

- Identification of the problem to be studied (the question and potential hypothesis to answer the question)
- Data collection (in the field or using an existing data set)
- Data preparation and organization
- Identification of response variables
- Identification of predictive variables (fixed factors)

- Identification of potential random factors (for mixed models)
- Model selection (for example, stepwise and averaging model procedures, AIC criteria, etc.)
- Assessing the accuracy of the SDM or goodness of fit (capacity to explain the variance of data)
- Hierarchical partitioning analysis or other alternatives in order to explore the relative importance of each predictor (covariate) on the response variable

The use of SDMs can be especially suitable for conservation planning [62]. In this book, we propose a procedure combining predictive statistical tools as SDMs, with the use of ornithological data and geographical information system (GIS) approaches, in order to set a protocol adequate to study and monitor HNV farmlands (Figs. 4.3 and 4.4).

4.4 Other Useful Tools

4.4.1 MRT—Multiregression Tree Analysis

Multiregression tree (MRT) or simply regression tree (RT) analysis [66] is an approach to recursively partition predictor variables used in applications dealing with categorical or remote sensing data when working with continuous response variables [67] as in the study cases provided in this book.

"Unlike classical regression techniques for which the relationship between the response and predictors is pre-specified (for example, straight line, quadratic) and the test is performed to prove or disprove the relationship, MRT or RT assumes no such relationship. It is primarily a method of constructing a set of decision rules on the predictor variables [68]. The rules are constructed by recursively partitioning

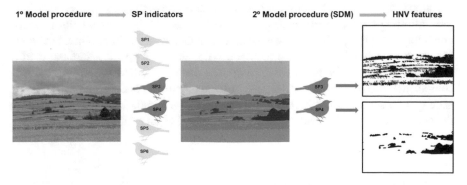

Fig. 4.3 Schematic representation of the proposed model procedure to select bird species as indicators of a given farming system (for instance, High Nature Value (HNV) farmland) (1° model procedure), and then application of a species distribution model (SDM) in order to identify the environmental characteristics of HNV farmlands supporting the occurrence of the bird indicators (2° model procedure)

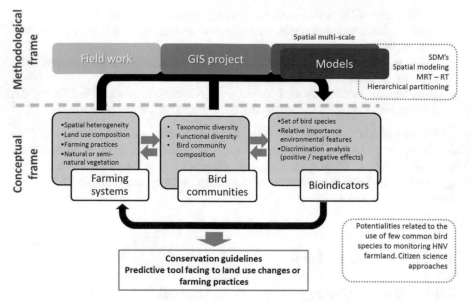

Fig. 4.4 Conceptual model of a proposed framework using bird species and a species distribution model (SDM) in order to study High Nature Value (HNV) farming areas

the data into successively smaller groups with binary splits based on a single predictor variable" [67].

MRT analysis is a suitable tool to be applied to data on bird communities, in order to study the threshold levels provided by the environmental characteristics of farming systems (for instance, HNV or non-HNV), affecting bird species richness or composition in farmland areas. The result of MRT is a tree whose 'leaves' (terminal groups of sites) are composed of subsets of sites chosen to minimize the within-group sums of squares (as in k-means clustering), but where each successive partition is defined by a threshold value or a state of one of the explanatory variables [69]. Multiregression tree analysis can be used to identify the critical values, where predictors are splitting the response variable in a dichotomous hierarchical manner. For example, when investigating the occurrence of a bird species 'indicator' of a particular kind of HNV farmland through the use of an SDM based on land use composition and landscape metrics, it should be possible to decide the proportion of each type of semi-natural feature it is necessary to maintain in order to guarantee the occurrence of the indicator species (see Fig. 4.5 as an example).

4.4.2   Indicator Species Analysis

Another useful tool, in order to identify a set of bird species as potential indicators in a particular type of farming system, is the indicator value or IndVal, which is a

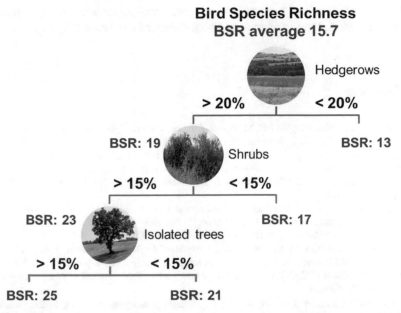

Bird Species Richness
BSR average 15.7

Fig. 4.5 Example of regression tree (RT) analysis outputs for bird species richness (BSR) in farmlands in Central Italy, based on a small sample site ($n = 50$). The explanatory variables are all the environmental variables collected in the field and related to the land use composition. Under each branch, the value in red (BSR) is the average species richness for the group. The leaves of the tree indicate the average catch weight per two of each species given the conditions and thresholds stipulated by the splits. The first split is stipulated on tree hedgerow coverage, the second one on shrub coverage and the third one on isolated tree coverage

hierarchical clustering method [70]. The IndVal can be used to select species that are typical from specific characteristics of the habitat or environment.

The formula used to compute the IndVal is:

$$IndVal_{ij} = Specificity_{ij} * Fidelity_{ij} * 100$$

where $IndVal_{ij}$ is the indicator value of an 'i' species in relation to a 'j' type of site, $Specificity_{ij}$ is the proportion of sites of type 'j' with species 'i' and, finally, $Fidelity_{ij}$ is the proportion of the number of individuals (abundance) of species 'i' that are in a 'j' type of site.

Any good indicator species would be those that are both abundant in a specific type of habitat (high specificity) and are predominantly found in this type of habitat (high fidelity).

The package 'indicspecies' from R is adequate in order to perform IndVal analysis [71], and then to calculate relationships between species and groups of sites (for instance, HNV farmlands versus non-HNV farmlands). Furthermore, new updates of this package provide a set of functions to assess the strength and statistical significance of the relationship between species occurrence/abundance and groups of sites, and even to improve indicator species analysis by combining groups of sites and group of species as indicators of site types [71, 72].

References

1. Bibby CJ, Hill DA, Burgess ND, Mustoe S. Bird census techniques. London: Academic; 2005.
2. Wuczyński A, Dajdok Z, Wierzcholska S, Kujawa K. Applying red lists to the evaluation of agricultural habitat: regular occurrence of threatened birds, vascular plants, and bryophytes in field margins of Poland. Biodivers Conserv [Internet]. Springer Netherlands; 2014 Apr 27 [cited 2016 Oct 12];23(4):999–1017. Available from: http://link.springer.com/10.1007/s10531-014-0649-y
3. Tryjanowski P, Morelli F. Presence of cuckoo reliably indicates high bird diversity: a case study in a farmland area. Ecol Indic [Internet]. Elsevier Ltd; 2015;55:52–8. Available from: http://linkinghub.elsevier.com/retrieve/pii/S1470160X15001363
4. Kosicki JZ, Zduniak P, Ostrowska M, Hromada M. Are predators negative or positive predictors of farmland bird species community on a large geographical scale? Ecol Indic [Internet]. 2016;62:259–70. Available from: http://linkinghub.elsevier.com/retrieve/pii/S1470160X15006342
5. Żmihorski M, Kotowska D, Berg Å, Pärt T. Evaluating conservation tools in Polish grasslands: the occurrence of birds in relation to agri-environment schemes and Natura 2000 areas. Biol Conserv [Internet]. 2016;194:150–7. Available from: http://linkinghub.elsevier.com/retrieve/pii/S0006320715301865
6. Surmacki A, Tryjanowski P. Efficiency of line transect and the point count methods in agricultural landscape of western Poland. Vogelwelt. 2000;120:201–4.
7. Budka M, Kokociński P. The efficiency of territory mapping, point-based censusing, and point-counting methods in censusing and monitoring a bird species with long-range acoustic communication—the corncrake Crex crex. Bird Study. Taylor & Francis; 2015;62(2):153–60.
8. Lindenmayer DB, Pierson J, Barton PS, Beger M, Branquinho C, Calhoun A, et al. A new framework for selecting environmental surrogates. Sci Total Environ [Internet]. Elsevier B. V.; 2015;538:1029–1038. Available from: http://linkinghub.elsevier.com/retrieve/pii/S0048969715305593
9. Rodrigues ASL, Brooks TM. Shortcuts for biodiversity conservation planning: the effectiveness of surrogates. Annu Rev Ecol Evol Syst. 2007;38(1):713–37.
10. Carignan V, Villard M-A. Selecting indicator species to monitor ecological integrity: a review. Environ Monit Assess [Internet]. Kluwer Academic Publishers; 2002 [cited 2016 Oct 12];78 (1):45–61. Available from: http://link.springer.com/10.1023/A:1016136723584
11. Grantham HS, Pressey RL, Wells JA, Beattie AJ. Effectiveness of biodiversity surrogates for conservation planning: different measures of effectiveness generate a kaleidoscope of variation. PLoS One [Internet]. 2010 Jan [cited 2014 May 23];5(7):e11430. Available from: http://www.pubmedcentral.nih.gov/articlerender.fcgi?artid=2904370&tool=pmcentrez&rendertype=abstract
12. Padoa-Schioppa E, Baietto M, Massa R, Bottoni L. Bird communities as bioindicators: the focal species concept in agricultural landscapes. Ecol Indic [Internet]. 2006 Jan [cited 2013 Jun 5];6(1):83–93. Available from: http://linkinghub.elsevier.com/retrieve/pii/S1470160X05000671
13. Carrascal LM, Cayuela L, Palomino D, Seoane J. What species-specific traits make a bird a better surrogate of native species richness? A test with insular avifauna. Biol Conserv [Internet]. 2012 Aug [cited 2014 Dec 14];152:204–11. Available from: http://www.sciencedirect.com/science/article/pii/S0006320712001917

14. Caro TM. Conservation by proxy. Indicator, umbrella, keystone, flagship, and other surrogate species. Washington, DC: Island Press; 2010. 400 p.
15. Billeter R, Liira J, Bailey D, Bugter R, Arens P, Augenstein I, et al. Indicators for biodiversity in agricultural landscapes: a pan-European study. J Appl Ecol [Internet]. 2007 Jul 23 [cited 2014 Jul 14];45(1):141–50. Available from: http://doi.wiley.com/10.1111/j.1365-2664.2007. 01393.x
16. Lambeck RJ. Focal species: a multi-species umbrella for nature conservation. Conserv Biol. 1997;11(4):849–56.
17. Noss RF. Indicators for monitoring biodiversity: a hierarchical approach. Conserv Biol [Internet]. 1990;4(4):355–64. Available from: http://doi.wiley.com/10.1111/j.1523-1739. 1990.tb00309.x
18. Burger J. Bioindicators: a review of their use in the environmental literature 1970–2005. Environ Bioindic [Internet]. Taylor & Francis; 2006 Jul [cited 2014 Apr 28];1(2):136–44. Available from: http://dx.doi.org/10.1080/15555270600701540
19. Holt EA, Miller SW. Bioindicators: using organisms to measure environmental impacts. Nat Educ Knowl. 2010;3(10):8.
20. Caro TM, O'Doherty G. On the use of surrogate species in conservation biology. Conserv Biol. 1999;13(4):805–14.
21. Loss SR, Ruiz MO, Brawn JD. Relationships between avian diversity, neighborhood age, income, and environmental characteristics of an urban landscape. Biol Conserv [Internet]. Elsevier Ltd; 2009 Nov [cited 2014 Jul 15];142(11):2578–85. Available from: http:// linkinghub.elsevier.com/retrieve/pii/S0006320709002596
22. Mikusinski G, Gromadzki M, Chylarecki P. Woodpeckers as indicators of bird diversity. Conserv Biol. 2001;15(1):208–17.
23. Herrando S, Weiserbs A, Quesada J, Ferrer X, Paquet JY. Development of urban bird indicators using data from monitoring schemes in two large European cities. Anim Biodivers Conserv. 2012;35(1):141–50.
24. Drever MC, Aitken KEH, Norris AR, Martin K. Woodpeckers as reliable indicators of bird richness, forest health and harvest. Biol Conserv [Internet]. 2008 Mar [cited 2014 Jun 10];141(3):624–34. Available from: http://www.sciencedirect.com/science/article/pii/ S0006320707004569
25. Morelli F, Jerzak L, Tryjanowski P. Birds as useful indicators of High Nature Value (HNV) farmland in Central Italy. Ecol Indic. 2014;38:236–42.
26. Noble D. The importance of indicators. Bird Popul. 2008;9:236–8.
27. Kosicki JZ, Chylarecki P. The hooded crow Corvus cornix density as a predictor of wetland bird species richness on a large geographical scale in Poland. Ecol Indic [Internet]. 2014 Mar [cited 2014 May 4];38:50–60. Available from: http://www.sciencedirect.com/science/article/ pii/S1470160X13004032
28. Gregory RD, van Strien A. Wild bird indicators: using composite population trends of birds as measures of environmental health. Ornithol Sci. 2010;22:3–22.
29. Skórka P, Mrtyka R, Wójcik JD. Species richness of breeding birds at a landscape scale— which habitat is the most important? Acta Ornithol. 2006;41(1):49–54.
30. Gregory RD, van Strien A, Voříšek P, Gmelig Meyling AW, Noble DG, Foppen RPB, et al. Developing indicators for European birds. Philos Trans R Soc London B Biol Sci [Internet]. 2005 Feb 28 [cited 2014 Sep 14];360(1454):269–88. Available from: http://www. pubmedcentral.nih.gov/articlerender.fcgi?artid=1569455&tool=pmcentrez& rendertype=abstract
31. Princé K, Lorrillière R, Barbet-Massin M, Jiguet F. Predicting the fate of French bird communities under agriculture and climate change scenarios. Environ Sci Policy [Internet]. 2013;33:120–32. Available from: http://linkinghub.elsevier.com/retrieve/pii/ S1462901113000944
32. Devictor V, Julliard R, Jiguet F. Distribution of specialist and generalist species along spatial gradients of habitat disturbance and fragmentation. Oikos. 2008;117(4):507–14.

33. Julliard R, Clavel J, Devictor V, Jiguet F, Couvet D. Spatial segregation of specialists and generalists in bird communities. Ecol Lett [Internet]. 2006 Nov [cited 2014 Jan 22];9 (11):1237–44. Available from: http://www.ncbi.nlm.nih.gov/pubmed/17040326

34. Reif J, Jiguet F, Šťastný K. Habitat specialization of birds in the Czech Republic: comparison of objective measures with expert opinion. Bird Study. 2010;57(2):197–212.

35. Devictor V, Julliard R, Clavel J, Jiguet F, Lee A, Couvet D. Functional biotic homogenization of bird communities in disturbed landscapes. Glob Ecol Biogeogr [Internet]. 2008 Mar [cited 2011 Jul 19];17(2):252–61. Available from: http://doi.wiley.com/10.1111/j.1466-8238.2007. 00364.x

36. Aue B, Diekötter T, Gottschalk TK, Wolters V, Hotes S. How High Nature Value (HNV) farmland is related to bird diversity in agro-ecosystems—towards a versatile tool for biodiversity monitoring and conservation planning. Agric Ecosyst Environ [Internet]. Elsevier B. V.; 2014;194:58–64. Available from: http://dx.doi.org/10.1016/j.agee.2014.04.012

37. Morelli F. Indicator species for avian biodiversity hotspots: combination of specialists and generalists is necessary in less natural environments. J Nat Conserv. 2015;27:54–62.

38. Kujawa K. Population density and species composition changes for breeding bird species in farmland woodlots in western Poland between 1964 and 1994. Agric Ecosyst Environ. 2002;91:261–71.

39. Laureto LMO, Cianciaruso MV, Samia DSM. Functional diversity: an overview of its history and applicability. Nat e Conserv [Internet]. Associação Brasileira de Ciência Ecológica e Conservação; 2015;13(2):112–6. Available from: http://dx.doi.org/10.1016/j.ncon.2015.11. 001

40. Luck GW, Carter A, Smallbone L. Changes in bird functional diversity across multiple land uses: interpretations of functional redundancy depend on functional group identity. PLoS One [Internet]. 2013 Jan [cited 2014 Jan 22];8(5):e63671. Available from: http://www.pubmedcentral. nih.gov/articlerender.fcgi?artid=3656964&tool=pmcentrez&rendertype=abstract

41. Guilhaumon F, Albouy C, Claudet J, Velez L, Ben Rais Lasram F, Tomasini J-A, et al. Representing taxonomic, phylogenetic and functional diversity: new challenges for Mediterranean marine-protected areas. Divers Distrib [Internet]. 2015;21(2):175–87. Available from: http://doi.wiley.com/10.1111/ddi.12280

42. Santini L, Belmaker J, Costello MJ, Pereira HM, Rossberg AG, Schipper AM, et al. Assessing the suitability of diversity metrics to detect biodiversity change. Biol Conserv. 2016

43. Tucker CM, Cadotte MW, Carvalho SB, Davies TJ, Ferrier S, Fritz SA, et al. A guide to phylogenetic metrics for conservation, community ecology and macroecology. Biol Rev [Internet]. 2016; Available from: http://doi.wiley.com/10.1111/brv.12252

44. Magurran A. Measuring biological diversity. Oxford: Blackwell Science; 2004.

45. Pearman PB, Lavergne S, Roquet C, Wüest R, Zimmermann NE, Thuiller W. Phylogenetic patterns of climatic, habitat and trophic niches in a European avian assemblage. Glob Ecol Biogeogr [Internet]. 2014 Nov 4 [cited 2013 Nov 6];23(4):414–24. Available from: http://doi. wiley.com/10.1111/geb.12127

46. de Bello F, Lavorel S, Gerhold P, Reier Ü, Pärtel M. A biodiversity monitoring framework for practical conservation of grasslands and shrublands. Biol Conserv [Internet]. 2010 Jan [cited 2014 Sep 10];143(1):9–17. Available from: http://www.sciencedirect.com/science/article/pii/ S0006320709002055

47. Petchey OL, Gaston KJ. Functional diversity (FD), species richness and community composition. Ecol Lett [Internet]. 2002 May [cited 2014 Apr 28];5(3):402–11. Available from: http:// doi.wiley.com/10.1046/j.1461-0248.2002.00339.x

48. Petchey OL, Gaston KJ. Functional diversity: back to basics and looking forward. Ecol Lett [Internet]. 2006 Jun [cited 2014 Jul 9];9(6):741–58. Available from: http://www.ncbi.nlm.nih. gov/pubmed/16706917

49. Villéger S, Mason NWH, Mouillot D. New multidimensional functional diversity indices for a multifaceted framework in functional ecology. Ecology. 2008;89(8):2290–301.

50. Frishkoff LO, Karp DS, M'Gonigle LK, Mendenhall CD, Zook J, Kremen C, et al. Loss of avian phylogenetic diversity in neotropical agricultural systems. Science (80-) [Internet]. 2014 Sep 12 [cited 2014 Oct 17];345(6202):1343–6. Available from: http://www.ncbi.nlm.nih.gov/pubmed/25214627

51. Faith DP. Conservation evaluation and phylogenetic diversity. Biol Conserv [Internet]. 1992;61 (1):1–10. Available from: http://linkinghub.elsevier.com/retrieve/pii/0006320792912013

52. EDGE of Existence. www.edgeofexistence.org [Internet]. 2015. Available from: www. edgeofexistence.org

53. Meynard CN, Devictor V, Mouillot D, Thuiller W, Jiguet F, Mouquet N. Beyond taxonomic diversity patterns: how do α, β and γ components of bird functional and phylogenetic diversity respond to environmental gradients across France? Glob Ecol Biogeogr [Internet]. 2011 Feb 28 [cited 2011 Jun 20];20(6):893–903. Available from: http://doi.wiley.com/10.1111/j.1466-8238.2010.00647.x

54. Topping CJ, Odderskær P, Kahlert J, Butler S, Vickery J, Norris K, et al. Modelling skylarks (Alauda arvensis) to predict impacts of changes in land management and policy: development and testing of an agent-based model. Bolhuis JJ, editor. PLoS One [Internet]. Public Library of Science; 2013 Jun 6 [cited 2016 Oct 13];8(6):e65803. Available from: http://dx.plos.org/10.1371/journal.pone.0065803

55. Guisan A, Thuiller W. Predicting species distribution: offering more than simple habitat models. Ecol Lett [Internet]. 2005 Sep [cited 2011 Jul 17];8(9):993–1009. Available from: http://doi.wiley.com/10.1111/j.1461-0248.2005.00792.x

56. Parolo G, Rossi G, Ferrarini A. Toward improved species niche modelling: Arnica montana in the Alps as a case study. J Appl Ecol. 2008;45:1410–8.

57. Elith J, Graham CH, Anderson RP, Dudık M, Ferrier S, Guisan A, et al. Novel methods improve prediction of species' distributions from occurrence data. Ecography (Cop). 2006;29:129–51.

58. Araújo MB, Guisan A. Five (or so) challenges for species distribution modelling. J Biogeogr [Internet]. 2006 Oct [cited 2011 Jun 11];33(10):1677–88. Available from: http://doi.wiley.com/10.1111/j.1365-2699.2006.01584.x

59. Jokimäki J, Suhonen J, Jokimäki-Kaisanlahti M-L, Carbó-Ramírez P. Effects of urbanization on breeding birds in European towns: impacts of species traits. Urban Ecosyst [Internet]. 2014 Oct 24 [cited 2014 Oct 29]; Available from: http://link.springer.com/10.1007/s11252-014-0423-7

60. Gormley AM, Forsyth DM, Griffioen P, Lindeman M, Ramsey DS, Scroggie MP, et al. Using presence-only and presence-absence data to estimate the current and potential distributions of established invasive species. J Appl Ecol [Internet]. Wiley-Blackwell; 2011 Feb [cited 2016 Oct 13];48(1):25–34. Available from: http://www.ncbi.nlm.nih.gov/pubmed/21339812

61. Elith J, Leathwick JR. Species distribution models: ecological explanation and prediction across space and time. Annu Rev Ecol Evol Syst [Internet]. Annual Reviews; 2009 Dec 6 [cited 2013 Aug 6];40(1):677–97. Available from: http://www.annualreviews.org/doi/abs/10.1146/annurev.ecolsys.110308.120159

62. Guisan A, Tingley R, Baumgartner JB, Naujokaitis-Lewis I, Sutcliffe PR, Tulloch AIT, et al. Predicting species distributions for conservation decisions. Ecol Lett [Internet]. 2013 Oct 17 [cited 2013 Nov 6];16:1424–35. Available from: http://www.ncbi.nlm.nih.gov/pubmed/24134332

63. Guisan A, Zimmermann NE. Predictive habitat distribution models in ecology. Ecol Modell [Internet]. 2000;135(2–3):147–86. Available from: http://linkinghub.elsevier.com/retrieve/pii/S0304380000003549

64. Zimmermann NE, Edwards TC, Graham CH, Pearman PB, Svenning JC. New trends in species distribution modelling. Ecography (Cop) [Internet]. 2010 Dec 22 [cited 2014 Feb 20];33 (6):985–9. Available from: http://doi.wiley.com/10.1111/j.1600-0587.2010.06953.x

65. Franklin J. Mapping species distributions: spatial inference and prediction. Cambridge: Cambridge University Press; 1997. 340 p.

66. De'ath G. Multivariate regression trees: a new technique for modeling species–environment relationships. Ecology. 2002;83(4):1105–17.
67. Prasad AM, Iverson LR, Liaw A. Newer classification and regression tree techniques: bagging and random forests for ecological prediction. Ecosystems. 2006;9(2):181–99.
68. Breiman L, Freidman J, Olshen R, Stone C. Classification and regression trees. Belmont: Chapman & Hall; 1984. 368 p.
69. Borcard D, Gillet F, Legendre P. Numerical ecology with R [Internet]. Media. New York, NY: Springer New York; 2011 [cited 2014 Jul 14]. 1–306 p. Available from: http://www.mendeley.com/catalog/numerical-ecology-r/
70. Dufrene M, Legendre P. Species assemblages and indicator species: the need for a flexible asymmetrical approach. Ecol Monogr. 1997;67:345–66.
71. De Cáceres M, Jansen F. "indicspecies" R package—functions to assess the strength and significance of relationship of species site group associations. 2016.
72. De Cáceres M, Legendre P, Moretti M. Improving indicator species analysis by combining groups of sites. Oikos [Internet]. 2010 Oct 14 [cited 2014 Jan 21];119(10):1674–84. Available from: http://doi.wiley.com/10.1111/j.1600-0706.2010.18334.x

Part I
Case Studies

Chapter 5
Case Study 1. Bird as Indicators of HNV: Case Study in Farmlands from Central Italy

Federico Morelli, Leszek Jerzak, and Piotr Tryjanowski

Abstract

- Country: Italy
- Year of study: 2011
- Dominant environment in study area: croplands
- Ecoregion: Mediterranean north
- Climate: temperate
- Target indicator: set of common bird species
- Main statistical tools: GLM, GLMM, indicator analysis
- Some useful R packages: 'AICcmodavg' [1], 'MASS' [2], 'glmmADMB' [3, 4], 'lme4' [5], 'MuMIn' [6], 'indicspecies' [7]

Summary Case Study No. 1

- Country: Italy
- Year of study: 2011
- Dominant environment in study area: croplands
- Ecoregion: Mediterranean north
- Climate: temperate
- Target indicator: set of common bird species
- Main statistical tools: GLM, GLMM, indicator analysis
- Some useful R packages: 'AICcmodavg' [1], 'MASS' [2], 'glmmADMB' [3, 4], 'lme4' [5], 'MuMIn' [6], 'indicspecies' [7]

F. Morelli (✉)
Faculty of Environmental Sciences, Department of Applied Geoinformatics and Spatial Planning, Czech University of Life Sciences Prague, Kamýcká 129, CZ-165 00 Prague 6, Czech Republic
e-mail: fmorellius@gmail.com

L. Jerzak
Faculty of Biological Sciences, University of Zielona Góra, Prof. Z. Szafrana St. 1, PL 65-516 Zielona Góra, Poland

P. Tryjanowski
Institute of Zoology, Poznań University of Life Sciences, Wojska Polskiego 71 C, PL-60-625 Poznań, Poland

© Springer International Publishing AG 2017   71
F. Morelli, P. Tryjanowski (eds.), *Birds as Useful Indicators of High Nature Value Farmlands*, DOI 10.1007/978-3-319-50284-7_5

The need for measures to prevent the loss of—and to monitor, in real time, the quality of—High Nature Value (HNV) farmlands, and to mitigate the loss of biodiversity, is widely acknowledged but requires urgent attention. Some recent studies showed how agri-environmental measures adopted in the countries of the European Community may be ineffective in order to guarantee the stop on biodiversity in some intensive farmland [8–12]. However, more studies focusing on strategies to increase the effectiveness of agri-environmental measures for conservation are necessary [8, 13, 14].

This case study performed in 2014 in Central Italy [15] examined the relationships between HNV farmland and the occurrence of common bird species, by means of species distribution models (SDMs). This protocol was used to define a set of bird species useful as indicators of HNV farmland. Furthermore, the relative importance of the characteristic of farmland was examined in order to explain the distribution of the selected bird species indicators of HNV farmlands.

5.1 Methodology

5.1.1 Study Area

The study was carried out in an agricultural landscape of the Northeastern Marches region in Central Italy (43° 46′ N; 12° 42′ E) (Fig. 5.1) covering circa 65,000 ha, ranging from 0 to 350 m a.s.l., among low hills and the Adriatic coasts. The climate in Central Italy is temperate [16] and characterized by high spring and summer temperatures and a marked summer drought. This area was selected because it

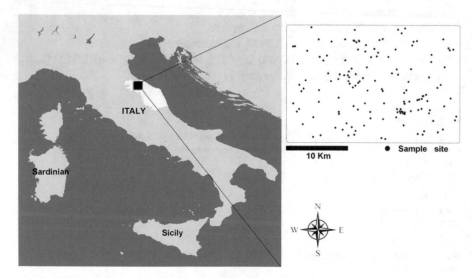

Fig. 5.1 Study area and sample sites in the Marches region, Central Italy

Fig. 5.2 Farming types in the Marches region, Central Italy: (**a–d**) some examples of High Nature Value farmland (Photo: F. Morelli, Y. Benedetti)

contains a large variety of different farming practices (see Figs. 5.2 and 5.3, photos). Within this area 160 sites were selected at random, separated by at least 500 m from each other.

5.1.2   Species and Environmental Data

In order to record the farmland bird communities, point counts were used, recording all the birds detected visually and acoustically within a radius of 100 m from the observer over a 10 min period [17]. The data were collected from 15 May to 30 June 2011 during the breeding season. Bird counts were carried out in the morning within 5 h following sunrise, only under good weather conditions. The sampled sites were visited at least twice during the breeding season. Non-breeding or raptor species were excluded from subsequent analysis as they require different survey methods to be collected.

In each sampled site the species richness was calculated as the maximum number of birds recorded. The species frequency was calculated as the percentage of occurrence on the total sampled points. From the 55 bird species recorded, 25% are listed on the national red list as vulnerable species. The remaining species are

Fig. 5.3 Farming types in the Marches region, Central Italy: (**a–d**) some examples of non-High Nature Value farmland (Photo: F. Morelli, Y. Benedetti)

listed in categories of lesser concern (near threatened, of least concern, data deficient or not evaluated) [18]. Fifty-four percent of recorded species are considered of European conservation concern (SPEC1, SPEC2, SPEC3) [19].

Habitat descriptions of a 100 m radius area around the sampled point were made in order to quantify the land use composition and structural characteristics of each site, i.e. covering 3 ha each. The percentages of land use composition and all semi-natural and marginal vegetation typologies within these areas were calculated through geographical information system (GIS) analysis (use of an intersect operator between the regional land cover map 1:10,000 [20] and buffer areas generated around the sampled points (using as the centre the coordinates from point counts)) (Fig. 5.4).

Furthermore, three different landscape metrics were calculated in each farmland. The complete list of environmental parameters is shown in Table 5.1, classified as land use, landscape or landscape metric type. The land use diversity at each site was calculated using the Shannon-Weaver diversity index, $H' = -\Sigma p_i \times \log p_i$, where p_i is the relative proportion of land use $_i$). The edge density was calculated as the sum of the perimeters of all polygons in the buffer zone [21]. The land use evenness index was calculated by dividing the Shannon diversity index by its maximum. This index ranged between 0 and 1 and is relatively easy to interpret.

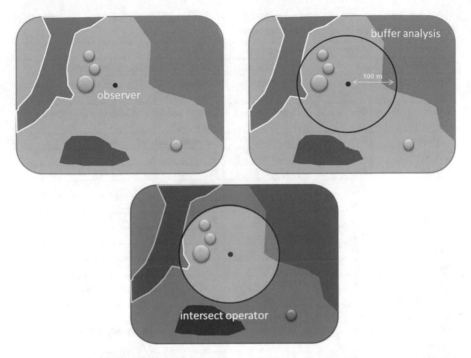

Fig. 5.4 Buffer analysis and use of an intersect operator to calculate the land use composition and landscape metrics in each plot

All farmland sites were classified, using a binary variable, as HNV [1] or non-HNV (0) farmland, following the classification applied by Galdenzi et al. [22]. The classification was derived by overlapping Galdenzi's map of HNV in the Marches region of Central Italy with our farmland sites. Sites were also classified as HNV if they had values of 'mv' (marginal vegetation, Table 5.1) and 'lud' above the mean for all 160 sites. An indication of 'extensive' farming practice was collected in situ during the surveys or from the presence of grasslands. Unfortunately, data on fertilizer and pesticide applications were not available.

5.1.3 Data Analysis

1. Preliminary comparisons

Difference in bird species richness in different farmland categories (HNV and non-HNV) was tested using a student's *t* test.

2. General model structure

The nature and strength of relationships between HNV farmland and bird species occurrence were examined using generalized linear models (GLMs) [23] with a

Table 5.1 Environmental descriptors of farmland in Central Italy. The spatial scale where metrics were measured was 3 ha around the point count

Parameter	Abbreviation	Level	Details
Altitude	alt	Landscape	Altitude of the point count (m/a.s.l.)
Slope	slo	Landscape	No slope (<3°): 0; slight slope (3–8°): 1; slope (>8°): 2
Roads	roa	Landscape	Length and type of roads (paved and unpaved)
Powerlines	pow	Landscape	Number of electricity wires
Urban	urb	Land use	%
Forest	for	Land use	%
Shrubs	shr	Land use	%
Uncultivated	unc	Land use	%
Badland	bad	Land use	%
Grassland	gra	Land use	%
Hedgerows	hed	Land use	%
Isolated trees	tre	Land use	%
Vineyards	vin	Land use	%
Olive orchards	oli	Land use	%
Cultivated	cul	Land use	%
Marginal vegetation	mv	Land use	Sum of coverage on shr, unc, hed, tre
Land use diversity	lud	Landscape metric	Shannon diversity index on land uses
Edge density	edg	Landscape metric	Perimeters of all polygons in the buffer
Land use evenness	eve	Landscape metric	Evenness index on land uses

dependent variable (HNV) modelled specifying a binomial distribution. Independent variables (predictors) were expressed as categorical variables: the presence [1] and absence (0) of each bird species.

From the total bird species recorded, for statistical purposes, only breeding bird species with frequencies of occurrence ≥10% over all sampled sites were selected. Furthermore, to avoid multicollinearity issues and to reduce the number of independent variables, avoiding also overfitting problems in models [24], parameters with the strongest correlation between them (>0.7) were manually removed. The full model was performed using 21 bird species as predictors. A stepwise backwards procedure was followed in order to select the best predictors using the Akaike information criterion (AIC) [25]. The bird species selected in the best model were considered the better predictors or indicators of HNV farmland.

3. Spatial autocorrelation issues

Sites were treated as independent units because the spatial autocorrelation (SAC) between geographical distance and species presence on sites was not significant (Mantel test $r = 0.203$, $n = 160$ sites, $p > 0.05$) [26, 27]. Spatial autocorrelation (SAC) analysis was based on the geographical distance matrix and the matrix of differences in bird species richness among sites, applying Monte Carlo permutations with 999 randomizations to test for significance [28].

4. Goodness-of-fit measures

The goodness of fit (GOF) of the statistical models describes how well they fit a set of observations, summarizing the discrepancy between the observed values and the values expected under a statistical model. In this work, we used the area under the receiver operating characteristic (ROC) curve (AUC) in order to measure the GOF of the best SDMs. The ROC is a graphic plot that illustrates the performance of a binary classifier system as its discrimination threshold is varied [29]. It is created by plotting the fraction of true positives out of the positives (true positive rate) versus the fraction of false positives out of the negatives (false positive rate), at various threshold settings [30]. The AUC calculated for each SDM selecting the predictive performance is expressed as an index ranging from 0.5 to 1 [31]. An approximate guide for classifying the accuracy of the AUC is that proposed by Swets [32]: 0.90–1.00 excellent; 0.80–0.90 good; 0.70–0.80 fair; 0.60–0.70 poor; 0.50–0.60 fail.

5. Species distribution models on selected predictors

We performed a series of SDMs [33] using separately the occurrence of each bird species selected as indicators of HNV as the response variable, and the environmental characteristics of the farmlands as independent variables (predictors). Independent predictive variables were transformed using the arcsin root square in the case of proportions.

Finally, the total contribution of each environmental parameter was calculated by means of the hierarchical partitioning protocol [34] on the series of SDMs. The relative importance of independent variables was measured using the MuMIn package in R, which employs GOF for each possible model to identify the variables that mostly affect the dependent variable [6]. The output of the relative importance for each predictor variable is a numeric vector ranging from 0 to 1, resulting from the sum of the 'Akaike weights' over all models including the explanatory variable [35].

All tests were performed with the R programme [36].

5.2   Results

5.2.1   Farmland Classification and Description

Of the 160 farmland sites surveyed, 34% were classified as HNV farmlands. The most common road type in the farmland sites was paved (in 67% of cases), while

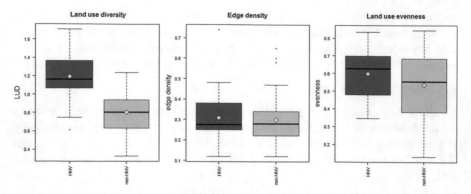

Fig. 5.5 Comparison between some landscape metrics values in High Nature Value (HNV) and non-HNV farmlands in Central Italy. The *boxplots* show the median (*black bar* in the middle of the *rectangle*), mean (*yellow rhombus*), upper and lower quartiles and extreme values

24% of sites had unpaved roads, and 9% were mixed (pavement and unpaved). Power lines were widespread on farmland (at 90% of sites). The most common slope category was 'no slope' or 'slight slope', with only 10% of sites being steeper. HNV farmlands showed higher values on the landscape metrics (Fig. 5.5).

Bird species richness was significantly greater in HNV farmlands (mean ± SD 16.8 ± 3.1, range 12–25) than in non-HNV farmlands (mean 14.5 ± SD 4.1, range 5–22) ($t = -3.78$, df 158, $p < 0.001$).

5.2.2 Bird Indicators of HNV Farmland

The best model using bird species as predictors of HNV farmland was obtained using only seven bird species, six of which were statistically significant for predicting HNV farmland (four positively and two negatively; Table 5.2). The best model was very highly performing, with a GOF value of 0.88, indicating that about 88% of cases were classified correctly by the model, approaching good–excellent performance [32]. The model using only four bird species positively correlated with HNV (Fig. 5.6) showed performance only slightly lower (0.82).

5.2.3 Relative Importance of HNV Farmland Characteristics for Bird Distribution

The most important characteristics of farmland for the distribution of the bird species selected as indicators of HNV were hedgerows (present in >86% of best

Table 5.2 Frequency, International Union for Conservation of Nature (IUCN) status, national conservation status [18], generalist/specialist species classification and results of logistic regression for the best model relating High Nature Value (HNV) farmland to bird species occurrence in Central Italy

Bird species	Name	Freq. (%)	IUCN RL	Italian RL	Spec	Estimate	SE	Z	p value	Rel. imp.
T. merula	Blackbird	76.3	LC	LC	G	3.79	1.22	3.09	**0.002****	0.995
S. communis	Common whitethroat	24.5	LC	LC	S	1.76	0.64	2.76	**0.006****	0.967
M. calandra	Corn bunting	32.5	LC	LC	S	2.43	0.93	2.62	**0.008***	0.776
P. dom. italiae	Italian sparrow	76.3	–	VU	G	1.29	0.62	2.07	**0.038***	0.568
C. chloris	European greenfinch	37.5	LC	NT	G	−2.54	0.62	−4.07	**0.000*****	0.998
A. arvensis	Sky lark	33.7	LC	VU	S	−1.31	0.65	−2.02	**0.044***	0.75
L. collurio	Red-backed shrike	16.3	LC	VU	S	−1.37	0.76	−1.81	0.070	0.477

The table shows the significant parameters selected after a stepwise procedure using the Akaike Information Criterion (AIC) (significant p values in bold) on the multiple variables analysis. The first four rows are bird species that are positively related to HNV farmland. The next two are negatively related to HNV farmland. The AIC value of the best model was 147.32, lowest if compared with the AIC value for the full model (163.94). Goodness of fit of the model: 88%. Significance levels: * $P \leq 0.05$; ** $P \leq 0.01$; *** $P \leq 0.001$

RL red list category, LC least concern, NT near threatened, VU vulnerable, Spec classification of generalist (G) and specialist (S) bird species [following the classification of the British Trust for Ornithology (BTO) and the Royal Society for the Protection of Birds (RSPB), available online (http://www.bto.org/)], SE standard error, Rel. Imp. relative importance of each predictor

Fig. 5.6 Four common bird species correlated with High Nature Value (HNV) farmlands in Central Italy. (**a**) *Passer domesticus italiae*; (**b**) *Milaria calandra*; (**c**) *Turdus merula*; (**d**) *Sylvia communis* (Photo (**a**): F. Pruscini; photos (**b–d**): C. Korkosz)

models, highly significantly three times ($p < 0.05$) and three times significantly at $p < 0.1$), followed by shrubs, isolated trees, uncultivated patches and land use diversity (all 43%) and grassland (29%). However, the relative importance of each HNV characteristic was different for each bird species (Table 5.3). The occurrence of the four bird species positively correlated with HNV farmland was mainly explained by the cover of hedgerows and shrubs. On the other hand, the occurrence of the two bird species negatively correlated with HNV farmland was mainly explained by the cover of hedgerows and trees (avoided by *Alauda arvensis*) and land use diversity and grasslands (avoided by *Carduelis chloris*).

Furthermore, some bird species (e.g. *Carduelis chloris* and *Passer domesticus italiae*) were related positively or negatively to several environmental characteristics (four parameters for each), while *Milaria calandra* was characterized by one main variable to which it was strongly positively linked (shrubs) (Table 5.3).

Table 5.3 Relative importance of each characteristic of High Nature Value (HNV) farmland for bird occurrence in Central Italy

Species	shr	hed	tre	unc	gra	lud
T. merula	(+) 0.300	(+) **0.998*****	(−) 0.334	(−) 0.397	(−) 0.260	(+) **0.998*****
S. communis	(+) **0.937****	(+) **0.783**†	(+) 0.259	(+) **0.985****	(−) 0.263	(+) 0.357
M. calandra	(+) **0.999*****	(+) **0.665**†	(−) 0.287	(+) 0.363	(+) 0.365	(+) 0.500
P. dom. Italiae	(−) 0.365	(−) 0.264	(+) **0.920***	(−) **0.994****	(+) **0.851***	(−) **0.985****
C. chloris	(+) 0.255	(+) **0.605**†	(+) **0.969****	(+) **0.943****	(−) 0.292	(−) **0.999*****
A. arvensis	(+) 0.318	(−) **0.978***	(−) **0.714**†	(−) 0.262	(+) **0.274***	(−) 0.283
L. collurio	(+) **0.820***	(+) **0.586***	(+) 0.271	(+) 0.270	(−) 0.318	(−) 0.433

The importance values shown in the table are the Akaike weights obtained from the best model for each bird species. The significant parameters resulting in the best model for each bird species are identified by means of bold text and significance codes (***$p \leq 0.001$, **$p \leq 0.01$, *$p \leq 0.05$, †$p \leq 0.1$). In brackets are the signs of the model-averaged parameter estimates for the effects of environmental parameters on each bird species
shr shrubs, *hed* hedgerows, *tre* isolated trees, *unc* uncultivated patches, *gra* grassland, *lud* land use diversity

5.3   Discussion

HNV farmlands in Central Italy were relatively widely dispersed (34% of the studied farmland sites). Bird species richness was greater in HNV farmlands compared with standard farmlands, indicating that this parameter can be used as a surrogate of biodiversity [37–39]. However, it is important to be cautious with ecological interpretations, since the higher species richness recorded in the HNV farms may be due to an increase in the number of habitat-generalist species. Thus, further analysis should take into account the composition of the bird community, in terms of farmland specialists and habitat generalists, in order to best describe the variations in functional diversity. Other studies in Western Europe found that species richness was not higher within HNV farmland, but bird communities were composed of more specialist species than those in non-HNV areas [40].

In this study, six bird species were significantly correlated (four positively and two negatively) with HNV farmland, and it is possible to group these species ecologically as follows: species typical of urban green areas, hedgerows and forests (European greenfinch, *Carduelis chloris* (here the result was negatively related to HNV), and blackbird, *Turdus merula* (positively related to HNV in this study)) [41], species typical of shrublands and uncultivated areas (common whitethroat, *Sylvia communis*, and corn bunting, *Milaria calandra*) [41, 42], species typical of grassland and lowland arable areas (sky lark, *Alauda arvensis*) [43] and species more ubiquitous (Italian sparrow, *Passer domesticus italiae*) [44]. The last bird species excluded in the backwards selection procedures for the model (red-backed shrike, *Lanius collurio*) can be considered typical of heterogeneous or mixed farmland [45–48]. Two selected species are considered vulnerable and of national concern (red list), while the other species are considered to be of least concern [18].

The presence of just four bird species on a farmland was enough to classify it as HNV with a performance of 82%, and the contemporary absence of two other bird species increased the accuracy of our predictions to 88%.

The occurrence of each bird species used in the indicator, however, was related to different parameters of farmland. For example, *Milaria calandra* was strongly associated with a few parameters relating to 'field margins' (shrubs and hedgerows). Thus modifications to this parameter, such as making more big field margins on farmlands or preserving a few patches of shrubs near croplands, could be highly beneficial for this species, as has also been highlighted in other studies [49]. Conversely, several bird species were strongly related to different parameters, such as *Carduelis chloris* or *Passer domesticus italiae* (four out of six parameters), and such species are less susceptible to changes to any one of them. *Alauda arvensis*, in contrast, was negatively correlated with hedgerows but positively related to grassland patches.

The negative correlation between *Carduelis chloris* and HNV farmlands could be explained by the strong preference of the species for forestal patches (present only as marginal habitats in HNV farmlands) or urban tissues (virtually absent or present with very low coverage in HNV farmlands). In the case of *Alauda arvensis*, the negative correlation between this species and HNV farmlands could be explained by the fact that the agricultural landscapes monitored in our study were mainly composed of croplands and a few grasslands.

The importance for farmland bird diversity of marginal elements, considered unproductive elements of agricultural landscapes, is already known [50–54].

Our results underlined the relative importance of hedgerows (for four of the six selected species, the presence of hedgerows significantly positively explained their distribution; for one other species it was significantly negatively related), followed by shrubs and by surrogates of landscape heterogeneity (land use diversity). In fact, the presence of natural and semi-natural features, such as field margins or hedgerows, can increase the number of ecological niches available for several bird species, having a positive effect on bird species richness [55–59]. For several carnivorous passerines (insectivorous or predators of micromammals) and diurnal raptors, these residual elements on road margins can represent an optimal feeding habitat, because they constitute effective ecological corridors for wildlife [60, 61]. Our study also highlights how other land use categories such as shrubs, uncultivated land and abandoned land (all descriptors of HNV farmland) are important to explain the distribution of several bird species [48, 49, 52, 62].

Our results underline how while some studies have correlated HNV farmlands mainly with specialist or threatened species, the better set of bird species suitable to predict HNV farmlands in Central Italy is a set of a few relatively common bird species, three specialist species and three generalist species. The percentage is repeated also if we analyse only the four species positively correlated with HNV farmlands. And this fact can be interpreted as a confirmation of the importance that these types of farmlands represent also for more common bird species.

5.3.1 Utility of the Proposed Methodology

In summary, the use of the joint occurrence of common bird species (presence–absence) in a modelling approach based on SDMs is suitable not only to identify important areas for birds or biodiversity hotspots [63–65] but also to study and classify agro-ecosystems. The proposed framework seems to work well even on relatively medium-sized data sets. Anyway, we suggest also the use of a multimodel inference (MMI) approach when working on models with many covariates, considering what has been said by Stokes et al. [24] about the issues related to models with too many regressors and the overfitting risk. The data set should be at least 5–10 observations for each parameter considered in the full model. Otherwise, MMI could be a good statistical choice.

Furthermore, the use of the right parameters, in addition to the hierarchical partitioning protocol, constitutes a powerful tool to study and monitor HNV farmland from a conservationist perspective. As suggested by Doxa et al. [66], to some extent it is possible to reverse biodiversity decline caused by agricultural intensification, if appropriate management actions are taken in the near future on HNV farmland. For this reason, identification and study of a set of a few bird species suitable for assessment of HNV farmland could be used also as a tool for quick monitoring of HNV status in time, being useful as an 'alarm bell', for example, when one of these bird species disappears.

Clearly the set of bird species will be subject to changes following the local bird species and the characteristics of the area where the method will be applied. But the procedure could be also applied on different type of agro-ecosystems, taking into account these local differences.

In addition, knowledge of the characteristics of a landscape or environment that explains the occurrence of each bioindicator can constitute an address for conservation efforts, showing the target to be preserved—elements that may potentially buffer the effects of intensification on biodiversity. Mainly, the results of local-scale studies could provide high-quality data with improved precision, which could be useful from the applied point of view [67, 68]. This could be a useful device for ecological restoration planning at the local scale [69] in the slipstream of European agricultural policy. This is especially important considering that one of the main aims of the Common Agricultural Policy (CAP) is a reduction in the negative trends in biodiversity, entrusting each European country with the specific task of biodiversity conservation within its own territory [70–72].

The set of common birds resulting as indicators of HNV farmland in this study should be included in development of the Italian farmland bird index (FBI) and other indices of agricultural biodiversity [73–75] at a local or regional scale. Furthermore, the use of a set of a few common bird species, easily recognized in nature during surveys, opens new possibilities in conservation planning, providing a potential tool to obtain information by using citizen science or citizen programmes [76–78]. And because much of the work to conserve biodiversity is carried out by non-governmental conservation organizations with public support, the use of a set of a few common species is expected to be a valuable tool for monitoring temporal changes in HNV farmlands, involving also common citizens.

References

1. Mazerolle MJ. AICcmodavg: model selection and multimodel inference based on (Q)AIC(c). R package [Internet]. 2016. Available from: http://cran.r-project.org/package=AICcmodavg.
2. Venables WN, Ripley BD. Modern applied statistics with S. 4th ed. New York: Springer; 2002.
3. Skaug H, Fournier D, Nielsen A. glmmADMB: generalized linear mixed models using AD model builder—R package. 2013.
4. Fournier DA, Skaug HJ, Ancheta J, Ianelli J, Magnusson A, Maunder M, et al. AD Model builder: using automatic differentiation for statistical inference of highly parameterized complex nonlinear models. Optim Methods Softw. 2012;27:233–49.
5. Bates D, Maechler M, Bolker B, Walker S. lme4: linear mixed-effects models using Eigen and S4—R package [Internet]. 2014. Available from: http://cran.r-project.org/package=lme4
6. Bartoń K. MuMIn: multi-model inference, R package. 2013.
7. De Cáceres M, Jansen F. "indicspecies" R package—functions to assess the strength and significance of relationship of species site group associations. 2016.
8. Kleijn D, Sutherland WJ. How effective are European agri-environment schemes in conserving and promoting biodiversity? J Appl Ecol [Internet]. 2003 Dec [cited 2014 Apr 28];40 (6):947–69. Available from: http://doi.wiley.com/10.1111/j.1365-2664.2003.00868.x
9. Batáry P, Dicks LV, Kleijn D, Sutherland WJ. The role of agri-environment schemes in conservation and environmental management. Conserv Biol [Internet]. 2015;29(4):1006–16. Available from: http://doi.wiley.com/10.1111/cobi.12536
10. Kleijn D, Kohler F, Báldi A, Batáry P, Concepción ED, Clough Y, et al. On the relationship between farmland biodiversity and land-use intensity in Europe. Proc R Soc London B Biol Sci [Internet]. 2009 Mar 7 [cited 2011 Jul 16];276:903–9. Available from: http://www. pubmedcentral.nih.gov/articlerender.fcgi?artid=2664376&tool=pmcentrez&rendertype=abstract
11. Ohl C, Drechsler M, Johst K, Wätzold F. Compensation payments for habitat heterogeneity: existence, efficiency, and fairness considerations. Ecol Econ. 2008;67(2):162–74.
12. Konvicka M, Beneš J, Cizek O, Kopecek F, Konvicka O, Vitaz L. How too much care kills species: grassland reserves, agri-environmental schemes and extinction of *Colias myrmidone* (Lepidoptera: Pieridae) from its former stronghold. J Insect Conserv [Internet]. Springer Netherlands; 2008 Oct 5 [cited 2016 Jul 19];12(5):519–25. Available from: http://link. springer.com/10.1007/s10841-007-9092-7
13. Lomba A, Guerra C, Alonso J, Honrado JP, Jongman RHG, McCracken D. Mapping and monitoring High Nature Value farmlands: challenges in European landscapes. J Environ Manage [Internet]. Elsevier Ltd; 2014;143:140–50. Available from: http://dx.doi.org/10. 1016/j.jenvman.2014.04.029
14. Lomba A, Alves P, Jongman RHG, Mccracken DI. Reconciling nature conservation and traditional farming practices: a spatially explicit framework to assess the extent of High Nature Value farmlands in the European countryside. Ecol Evol. 2015;1–14.
15. Morelli F, Jerzak L, Tryjanowski P. Birds as useful indicators of High Nature Value (HNV) farmland in Central Italy. Ecol Indic. 2014;38:236–42.
16. Tomaselli R, Balduzzi A, Filipello S. Carta bioclimatica d'Italia. Scala 1:2.000.000. Istituto di Botanica – Università di Pavia. Ministero Agricoltura e Foreste; 1972.
17. Bibby CJ, Burgess ND, Hill DA. Bird census techniques (Google eBook) [Internet]. Academic Press; 1992 [cited 2014 Apr 24]. 257 p. Available from: http://books.google.com/books? id=5TqfwEHCVuoC&pgis=1

18. Peronace V, Cecere JG, Gustin M, Rondinini C. Lista rossa 2011 degli uccelli nidificanti in Italia. Avocetta. 2012;36(1):11–58.
19. BirdLife International. The BirdLife checklist of the birds of the world, with conservation status and taxonomic sources [Internet]. 2010. Available from: http://www.birdlife.info/docs/ SpcChecklist/C
20. AA.VV. Land-use map 1:10,000 Marche Region. Regional Cartographic Office, Ancona, Italy; 2010.
21. Šímová P, Gdulová K. Landscape indices behavior: a review of scale effects. Appl Geogr [Internet]. 2012 May [cited 2014 Sep 2];34(3):385–94. Available from: http://linkinghub. elsevier.com/retrieve/pii/S0143622812000057
22. Galdenzi D, Pesaresi S, Casavecchia S, Zivkovic L, Biondi E. The phytosociological and syndynamical mapping for the identification of High Nature Value farmland. Plant Sociol. 2012;49(2):59–69.
23. McCullagh P, Nelder JA. Generalized linear models. London: Chapman & Hall; 1989. 261 p.
24. Stokes ME, Davis CS, Koch GG. In: Cary N, editor. Categorical data analysis using the SAS system. 2nd ed. Cary: SAS Publishing, BBU Press and John Wiley Sons Inc.; 2009.
25. Akaike H. A new look at the statistical model identification. IEEE Trans Automat Contr [Internet]. IEEE; 1974 Dec 1 [cited 2014 Jul 12];19(6):716–23. Available from: http:// ieeexplore.ieee.org/articleDetails.jsp?arnumber=1100705
26. Guillot G. On the use of the simple and partial Mantel tests in presence of spatial auto-correlation. Math Model [Internet]. 2011;2011:1–13. Available from: http://arxiv.org/abs/ 1112.0651
27. Dormann CF, McPherson JM, Araújo MB, Bivand R, Bolliger J, Carl G, et al. Methods to account for spatial autocorrelation in the analysis of species distributional data: a review. Ecography (Cop) [Internet]. 2007 Oct 27 [cited 2014 Jul 10];30(5):609–28. Available from: http://doi.wiley.com/10.1111/j.2007.0906-7590.05171.x
28. Oksanen J, Guillaume Blanchet F, Kindt R, Legendre P, Minchin PR, O'Hara BR, et al. Vegan: community ecology package. R package version 2.3–4 [Internet]. 2016. p. 291. Available from: https://cran.r-project.org/package=vegan
29. Yu T. ROCS: receiver operating characteristic surface for class-skewed high-throughput data. Xing Y, editor. PLoS One [Internet]. Public Library of Science; 2012 Jan [cited 2014 Sep 4];7(7):e40598. Available from: http://dx.plos.org/10.1371/journal.pone.0040598
30. Liu C, White M, Newell G. Measuring the accuracy of species distribution models: a review. In: 18th World IMACS/MODSIM Congress, Cairns, Australia 13–17 July 2009 [Internet]. 2009. p. 4241–7. Available from: http://mssanz.org.au/modsim09
31. DeLong ER, DeLong DM, Clarke-Pearson DL. Comparing the areas under two or more correlated receiver operating characteristic curves: a non parametric approach. Biometrics. 1988;44(3):837–45.
32. Swets JA. Measuring the accuracy of diagnostic systems. Science (80-). 1988;240(4857):1285–93.
33. Elith J, Leathwick JR. Species distribution models: ecological explanation and prediction across space and time. Annu Rev Ecol Evol Syst [Internet]. Annual Reviews; 2009 Dec 6 [cited 2013 Aug 6];40(1):677–97. Available from: http://www.annualreviews.org/doi/abs/ 10.1146/annurev.ecolsys.110308.120159
34. Mac NR. Regression and model-building in conservation biology, biogeography and ecology: the distinction between—and reconciliation of—"predictive" and "explanatory" models. Biodivers Conserv. 2000;9:655–71.
35. Burnham KP, Anderson DR. Model selection and multimodel inference: a practical information-theoretic approach [Internet]. 2nd ed. New York: Springer-Verlag; 2002. 488 p. Available from: http://books.google.com/books?hl=it&lr=&id=fT1Iu-h6E-oC&pgis=1
36. R Development Core Team. R: a language and environment for statistical computing [Internet]. Vienna, Austria: R Foundation for Statistical Computing, Vienna, Austria; 2016. Available from: http://www.r-project.org

37. Caro TM, O'Doherty G. On the use of surrogate species in conservation biology. Conserv Biol. 1999;13(4):805–14.
38. Lindenmayer DB, Pierson J, Barton PS, Beger M, Branquinho C, Calhoun A, et al. A new framework for selecting environmental surrogates. Sci Total Environ [Internet]. Elsevier B.V.; 2015;538:1029–1038. Available from: http://linkinghub.elsevier.com/retrieve/pii/S0048969715305593
39. Devictor V, Mouillot D, Meynard CN, Jiguet F, Thuiller W, Mouquet N. Spatial mismatch and congruence between taxonomic, phylogenetic and functional diversity: the need for integrative conservation strategies in a changing world. Ecol Lett [Internet]. 2010 Aug 1 [cited 2012 Oct 10];13(8):1030–40. Available from: http://www.ncbi.nlm.nih.gov/pubmed/20545736
40. Pointereau P, Doxa A, Coulon F, Jiguet F, Paracchini ML. Analysis of spatial and temporal variations of High Nature Value farmland and links with changes in bird populations: a study on France [Internet]. JRC Scientific and Technical Reports 2010. Available from: http://eln-fab.eu/uploads/Analysis of spatial and temporal variations of High Nature Value farmland and links with changes in bird populations a study on France.pdf
41. Cramp S, Perrins C. The birds of the western palearctic. Oxford: Oxford University Press; 1994.
42. Donald PF, Evans AD. Habitat selection by corn buntings *Miliaria calandra* in winter. Bird Study. 1994;41:199–210.
43. Erdős S, Báldi A, Batáry P. Nest site selection and breeding ecology of sky larks *Alauda arvensis* in Hungarian farmland. Bird Study. 2009;56:259–63.
44. Brichetti P, Rubolini D, Galeotti P, Fasola M. Recent declines in urban Italian sparrow *Passer (domesticus) italiae* populations in northern Italy. Ibis (Lond 1859). 2008;150:177–81.
45. Goławski A, Goławska S. Habitat preference in territories of the red-backed shrike *Lanius collurio* and their food richness in an extensive agriculture landscape. Acta Zool Acad Sci Hung. 2008;54(1):89–97.
46. Svendsen JK, Sell H, Bøcher PK, Svenning J. Habitat and nest site preferences of red-backed shrike (*Lanius collurio*) in Western Denmark. Ornis Fenn. 2015;92.
47. Morelli F, Santolini R, Sisti D. Breeding habitat of red-backed shrike *Lanius collurio* on farmland hilly areas of Central Italy: is functional heterogeneity one important key? Ethol Ecol Evol. 2012;24(2):127–39.
48. Girardello M, Morelli F. Modelling the environmental niche of a declining farmland bird species. Ital J Zool [Internet]. 2012;79(3):434–40. Available from: http://dx.doi.org/10.1080/11250003.2012.666572
49. Kosiński Z, Tryjanowski P. Habitat selection of breeding seed-eating passerines on farmland in Western Poland. Ekol. 2000;19:307–16.
50. Fahrig L, Baudry J, Brotons L, Burel FG, Crist TO, Fuller RJ, et al. Functional landscape heterogeneity and animal biodiversity in agricultural landscapes. Ecol Lett [Internet]. 2011 Feb [cited 2014 Jan 23];14(2):101–12. Available from: http://www.ncbi.nlm.nih.gov/pubmed/21087380
51. Titeux N, Henle K, Mihoub J-B, Regos A, Geijzendorffer IR, Cramer W, et al. Biodiversity scenarios neglect future land use change. Glob Chang Biol. 2016; (March).
52. Ceresa F, Bogliani G, Pedrini P, Brambilla M. The importance of key marginal habitat features for birds in farmland: an assessment of habitat preferences of red-backed shrikes *Lanius collurio* in the Italian Alps. Bird Study [Internet]. BTO; 2012 Aug 13 [cited 2015 Feb 15];59(3):327–34. Available from: http://www.tandfonline.com/doi/abs/10.1080/00063657.2012.676623#.VOSmWPmG_X4
53. Benton TG, Vickery J, Wilson JD. Farmland biodiversity: is habitat heterogeneity the key? Trends Ecol Evol [Internet]. 2003 Apr [cited 2011 Jul 16];18(4):182–8. Available from: http://linkinghub.elsevier.com/retrieve/pii/S0169534703000119
54. Morelli F. Relative importance of marginal vegetation (shrubs, hedgerows, isolated trees) surrogate of HNV farmland for bird species distribution in Central Italy. Ecol Eng [Internet].

Elsevier B.V.; 2013 Aug [cited 2013 May 16];57:261–6. Available from: http://linkinghub. elsevier.com/retrieve/pii/S0925857413001614

55. Batáry P, Matthiesen T, Tscharntke T. Landscape-moderated importance of hedges in conserving farmland bird diversity of organic vs. conventional croplands and grasslands. Biol Conserv [Internet]. 2010 Sep [cited 2011 Jul 7];143(9):2020–7. Available from: http://linkinghub.elsevier.com/retrieve/pii/S0006320710002193

56. Parish T, Lakhani KH, Sparks TH. Modelling the relationship between bird population variables and hedgerow and other field margin attributes. I. Species richness of winter, summer and breeding birds. J Appl Ecol [Internet]. 1994 Nov [cited 2016 Jul 24];31(4):764–75. Available from: http://www.jstor.org/stable/2404166?origin=crossref

57. Parish T, Lakhani KH, Sparks TH. Modelling the relationship between bird population variables and hedgerow, and other field margin attributes. II. Abundance of individual species and of groups of similar species. J Appl Ecol [Internet]. 1995 May [cited 2016 Jul 24];32 (2):362–71. Available from: http://www.jstor.org/stable/2405102?origin=crossref

58. Green RE, Osborne PE, Sears EJ. The distribution of passerine birds in hedgerows during the breeding season in relation to characteristics of the hedgerow and adjacent farmland. J Appl Ecol [Internet]. 1994 Nov [cited 2016 Jul 24];31(4):677–92. Available from: http://www.jstor. org/stable/2404158?origin=crossref

59. Hinsley S. The influence of hedge structure, management and landscape context on the value of hedgerows to birds: a review. J Environ Manage [Internet]. 2000 Sep [cited 2011 Sep 1];60 (1):33–49. Available from: http://linkinghub.elsevier.com/retrieve/pii/S0301479700903608

60. Vermeulen HJW, Opdam PFM. Effectiveness of roadside verges as dispersal corridors for small ground-dwelling animals: a simulation study. Landsc Urban Plan [Internet]. 1995 Feb [cited 2014 Aug 11];31(1–3):233–48. Available from: http://www.sciencedirect.com/science/article/pii/016920446940105OI

61. Vermeulen HJW. Corridor function of a road verge for dispersal of stenotopic heathland ground beetles Carabidae. Biol Conserv [Internet]. 1994 Jan [cited 2014 Aug 11];69(3):339–49. Available from: http://www.sciencedirect.com/science/article/pii/0006320794904332

62. Morelli F. Plasticity of habitat selection by red-backed shrikes (*Lanius collurio*) breeding in different landscapes. Wilson J Ornithol [Internet]. 2012;124(1):51–6. Available from: http://www.bioone.org/doi/abs/10.1676/11-103.1

63. Guillera-Arroita G, Lahoz-Monfort JJ, Elith J, Gordon A, Kujala H, Lentini P, et al. Is my species distribution model fit for purpose? Matching data and models to applications. Glob Ecol Biogeogr. 2015;24(3):276–92.

64. Morelli F. Indicator species for avian biodiversity hotspots: combination of specialists and generalists is necessary in less natural environments. J Nat Conserv. 2015;27:54–62.

65. Guisan A, Tingley R, Baumgartner JB, Naujokaitis-Lewis I, Sutcliffe PR, Tulloch AIT, et al. Predicting species distributions for conservation decisions. Ecol Lett [Internet]. 2013 Oct 17 [cited 2013 Nov 6];16:1424–35. Available from: http://www.ncbi.nlm.nih.gov/pubmed/24134332

66. Doxa A, Paracchini ML, Pointereau P, Devictor V, Jiguet F. Preventing biotic homogenization of farmland bird communities: the role of High Nature Value farmland. Agric Ecosyst Environ. 2012;148:83–8.

67. Voříšek P, Klvaňová A, Gregory RD, Aunins A, Chylarecki P, Crowe O, et al. The state of Europe's common birds. Prague: CSO/RSPB ; 2007.23 pp

68. Tryjanowski P, Hartel T, Báldi A, Szymański P, Tobółka M, Herzon I, et al. Conservation of farmland birds faces different challenges in Western and Central–Eastern Europe. Acta Ornithol [Internet]. 2011 Jun [cited 2013 Feb 19];46(1):1–12. Available from: http://www. bioone.org/doi/abs/10.3161/000164511X589857

69. Abensperg-Traun M, Wrbka T, Bieringer G, Hobbs R, Deininger F, Main BY, et al. Ecological restoration in the slipstream of agricultural policy in the old and new world. Agric Ecosyst Environ [Internet]. 2004 Aug [cited 2014 Apr 28];103(3):601–11. Available from: http://www. sciencedirect.com/science/article/pii/S0167880903003815

70. Baldock D, Beaufoy G, Bennett G, Clark J. Nature conservation and new directions in the EC Common Agricultural Policy. London: Institute for European Environmental Policy (IEEP); 1993.

71. de la Concha I. The Common Agricultural Policy and the role of rural development programmes in the conservation of steppe birds. In: Lynx Edicions & Centre Tecnològic Forestal de Catalunya B, editor. In Bota G, Morales MB, Mañosa S, Camprodon J, editors. Ecology and conservation of steppe-land birds. ;2005. pp. 141–68.

72. Oñate JJ. A reformed CAP? Opportunities and threats for the conservation of steppe-birds and the agri-environment. In: Lynx Edicions & Centre Tecnològic Forestal de Catalunya B, editor. In Bota G, Morales M B, Mañosa S, Camprodon J, editors. Ecology and conservation of steppe-land birds; 2005. pp. 141–68.

73. Billeter R, Liira J, Bailey D, Bugter R, Arens P, Augenstein I, et al. Indicators for biodiversity in agricultural landscapes: a pan-European study. J Appl Ecol [Internet]. 2007 Jul 23 [cited 2014 Jul 14];45(1):141–50. Available from: http://doi.wiley.com/10.1111/j.1365-2664.2007.01393.x

74. Gregory RD, van Strien A, Vořišek P, Gmelig Meyling AW, Noble DG, Foppen RPB, et al. Developing indicators for European birds. Philos Trans R Soc London B Biol Sci [Internet]. 2005 Feb 28 [cited 2014 Sep 14];360(1454):269–88. Available from: http://www.pubmedcentral.nih.gov/articlerender.fcgi?artid=1569455&tool=pmcentrez&rendertype=abstract

75. Campedelli T, Tellini Florenzano G, Sorace A, Fornasari L, Londi G, Mini L. Species selection to develop an Italian farmland bird index. Avocetta. 2009;33:87–91.

76. McCaffrey RE. using citizen science in urban bird studies. Urban Habitats [Internet]. 2005;3(1):70–86. Available from: http://www.urbanhabitats.org

77. Cooper CB, Dickinson J, Phillips T, Bonney R. Citizen science as a tool for conservation in residential ecosystems. Ecol Soc [Internet]. 2007;12(2):11. Available from: http://www.ecologyandsociety.org/vol12/iss2/art11/

78. Devictor V, Whittaker RJ, Beltrame C. Beyond scarcity: citizen science programmes as useful tools for conservation biogeography. Divers Distrib [Internet]. 2010 Apr 13 [cited 2014 Jul 9];16(3):354–62. Available from: http://doi.wiley.com/10.1111/j.1472-4642.2009.00615.x

Chapter 6
Case Study 2. Birds as Indicators of HNV: Case Study in Portuguese Cork Oak *Montados*

João E. Rabaça, Luísa Catarino, Pedro Pereira, António Luís, and Carlos Godinho

Abstract

- Country: Portugal
- Years of study: 2011–2012
- Dominant environment in study area: cork oak *montados*
- Ecoregion: Mediterranean (South and North)
- Climate: temperate
- Target indicator: ecological bird guilds based on habitat preference
- Main statistical tools: mixed models, information theoretical approach, data dredging
- Some useful R packages: MuMIn [1] and nlme [2]

J.E. Rabaça (✉)
ICAAM – Instituto de Ciências Agrárias e Ambientais Mediterrânicas, Universidade de Évora, Núcleo da Mitra, Ap. 94, 7002-554 Évora, Portugal

LabOr – Laboratório de Ornitologia, ICAAM Universidade de Évora, 7002-554 Évora, Portugal

Departamento de Biologia, Universidade de Évora, 7002-554 Évora, Portugal
e-mail: jrabaca@uevora.pt

L. Catarino
LabOr – Laboratório de Ornitologia, ICAAM Universidade de Évora, 7002-554 Évora, Portugal

P. Pereira • C. Godinho
ICAAM – Instituto de Ciências Agrárias e Ambientais Mediterrânicas, Universidade de Évora, Núcleo da Mitra, Ap. 94, 7002-554 Évora, Portugal

LabOr – Laboratório de Ornitologia, ICAAM Universidade de Évora, 7002-554 Évora, Portugal

A. Luís
Departamento de Biologia, Universidade de Aveiro, Campus de Santiago, 3810-193 Aveiro, Portugal

© Springer International Publishing AG 2017   89
F. Morelli, P. Tryjanowski (eds.), *Birds as Useful Indicators of High Nature Value Farmlands*, DOI 10.1007/978-3-319-50284-7_6

Case Summary Study No. 2

- Country: Portugal
- Years of study: 2011–2012
- Dominant environment in study area: cork oak *montados*
- Ecoregion: Mediterranean (South and North)
- Climate: temperate
- Target indicator: ecological bird guilds based on habitat preference
- Main statistical tools: mixed models, information theoretical approach, data dredging
- Some useful R packages: MuMIn [1] and nlme [2]

Montados form a heterogeneous landscape of wooded matrix dominated by cork and/or holm oak with open areas characterized by fuzzy boundaries [3]. *Montados* support high biological diversity associated with low-intensity management and landscape diversity provided by a continuous gradient of land cover [4–7]. Among other features this permits the classification of *montados* as a High Nature Value (HNV) system. We assessed the role of birds as HNV indicators for *montados* and tested several bird groups—farmland, edge, forest generalist and forest specialist species; and some universal indicators such as species conservation status, the Shannon diversity index and species richness [8–10]. Our study areas covered the North–South distribution of cork oak in Portugal, and we surveyed breeding bird communities across 117 sampling sites. In addition to variables related to management and sanitary status, we considered variables that characterize the landscape heterogeneity within the *montados*—trees and shrub density and richness of woody vegetation. Our results suggest that specific bird guilds can be used as HNV indicators of particular typologies of *montado*, and highlight the need to develop an indicator that could be transversally applied to all types of *montado*.

6.1   Methodology

6.1.1   Study Area

We sampled four areas covering the main distribution range of cork oak in Portugal (Fig. 6.1): the site of community importance (SCI) of Romeu (hereafter referred to as Romeu) with several private owners (7° 1′ to 7° 6′ W and 41° 33′ to 41° 28′ N), Companhia das Lezírias S.A. (Lezírias), a public ownership farm (8° 48′ W and 38° 50′ N), SCI Serra de Monfurado (Monfurado) (7° 40′ to 8° 16′ W and 38° 27′ to 38° 41′ N) and Serra de Grândola (Grândola) (8° 34′ to 8° 38′ W and 38° 9′ to 38° 8′ N), both with several private owners. These areas reflect the most common typologies of *montados*—half of the national distribution of cork oak has a tree coverage between 10% and 30%, a quarter between 30% and 50% and a quarter superior to 50% [11]. Our sampling sites represent these three categories in a similar

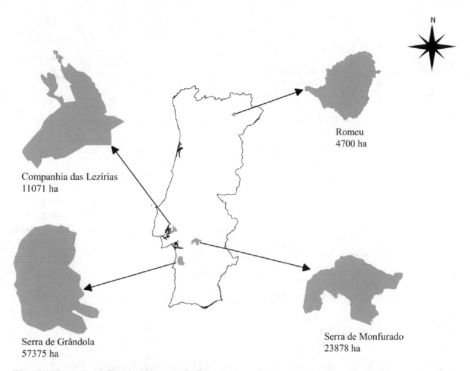

Fig. 6.1 Location of the study areas in Portugal

proportion. Lezírias and Monfurado had sites belonging to all categories; sites with higher tree cover were absent in Romeu and Grândola (Fig. 6.2). These areas are located in the Mediterranean part of the country (the Northeast and the entire Southern half of Portugal), characterized by hot and dry summers and moderate rainy winters. The altitude ranges between 15 m (Lezírias) and 600 m (Romeu). Romeu and Monfurado showed the lowest mean annual temperature (12.3 °C and 12.5 °C, respectively) and the highest levels of mean annual precipitation (760 and 800 mm, respectively). Lezírias and Grândola showed mean annual temperatures of 15.7 °C and 15.6 °C, respectively, and lower levels of mean annual precipitation (644 and 500 mm, correspondingly). The woodland area is dominated by cork oaks, but holm oak settlements can be found in Monfurado and mixed stands with cork oak, maritime pine (*Pinus pinaster*) and stone pine (*Pinus pinea*) occur in Lezírias. Other common land uses are rice fields, vineyards and pine woods in Lezírias; olive groves, small orchards, dry cereal fields and fallows in Monfurado; and olive groves and vineyards in Romeu. Riparian galleries are present in all areas.

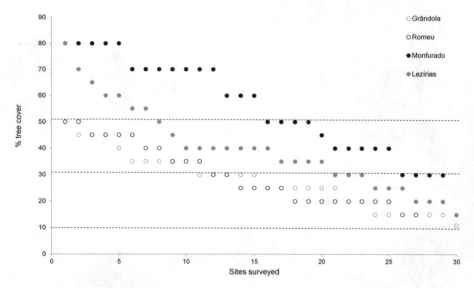

Fig. 6.2 Distribution of sampling sites along a gradient of *montado* coverage

6.1.2 Bird Census

We carried out bird censuses in Monfurado (2011) and in other areas (2012) during the breeding seasons (between April and May). Data on bird species were gathered using 10 min point counts (e.g. [12]) with a 100 m radius and surveys were conducted between 6:00 and 11:00 a.m. when birds are more active, and we avoided days with hard wind and rain. In each study area 30 sites were surveyed except in Romeu where we sampled 27 points. Our four areas covered a wide range of the different *montado* typologies present in Portugal (Fig. 6.2), and sites were randomly selected as long as they satisfied two criteria: [1] accessibility and [2] being situated at least 500 m apart from each other to minimize the probability of double counting birds. Three experienced observers with similar skills of bird detection conducted the surveys.

We excluded aerial-feeding birds (*e.g.* the barn swallow *Hirundo rustica*) from data treatment, as well as species with known large home ranges (*e.g.* carrion crows *Corvus corone* and ravens *C. corax*).

We used the species richness of each of four ecological guilds (farmland species, edge species, forest generalists and forest specialists) defined according to their habitat preferences in the Mediterranean part of Portugal during the breeding season [9] (Table 6.1). Species classification is related to the specialization degree, taking into account the habitats occupied along a gradient of natural terrestrial habitats, and may not be applicable outside this geographical area or time of year. This gradient can be represented by an increased complexity of vegetation structure, from structurally simpler habitats such as grasslands to intermediate habitats such as heathlands and scrublands, culminating in oak forests (such as cork oak or

Table 6.1 Ecological guilds based on species habitat preferences

Bird guild	Definition criteria	Species
Forest specialists	Species that occur only in certain forest types	*Dendrocopos minor, Troglodytes troglodytes, Erithacus rubecula, Phoenicurus phoenicurus, Sylvia atricapilla, Phylloscopus bonelli, Phylloscopus ibericus, Aegithalos caudatus, Sitta europaea, Coccothraustes coccothraustes*
Forest generalists	Forest species that occur in any woody areas, regardless of the density and height of the plant formations, as well as the bioclimatic conditions of the site	*Columba palumbus, Cuculus canorus, Dendrocopos major, Luscinia megarhynchos, Turdus merula, Sylvia melanocephala, Lophophanes cristatus, Cyanistes caeruleus, Parus major, Certhia brachydactyla, Garrulus glandarius, Oriolus oriolus, Fringilla coelebs, Serinus serinus, Carduelis chloris*
Edge species	Species that occur at the boundary between two or more habitats	*Streptopelia turtur, Clamator glandarius, Jynx torquilla, Lullula arborea, Hippolais polyglotta, Sylvia undata, Sylvia cantillans, Lanius meridionalis, Lanius senator, Passer montanus, Petronia petronia, Carduelis carduelis, Carduelis cannabina, Emberiza cirlus, Emberiza cia*
Farmland species	Farmland species that tolerate low tree densities	*Alectoris rufa, Coturnix coturnix, Upupa epops, Galerida theklae, Oenanthe hispanica, Saxicola rubicula, Cisticola juncidis, Sturnus unicolor, Passer domesticus, Emberiza calandra*

holm oak). To this end, we used the available information regarding places and strategies used to capture food, sites used for territorial defence and nesting sites. Farmland species mostly occur in association with open fields or, at most, with scattered shrubs or trees; edge species are associated with transitional areas that can occur at boundaries between two or more habitats (*e.g.* forest and open areas); forest generalists occur in any woody areas, regardless of the density and height of the tree formations; forest specialists tend to occupy stable forest environments and are therefore more sensitive to human disturbance than generalist forest species (some examples of species for each guild are shown in Fig. 6.3). Most occur in natural or semi-natural forests with the following features: [1] a high canopy of great maturity; [2] large stratification of woody vegetation; and [3] a humid microclimate.

In addition to ecological guilds we also considered the Shannon diversity index (ShInd), total species richness (SpRich) and a variable that takes into account the richness of species with conservation status (ConSt). This variable was estimated as the richness by a point count of species listed in at least one of the following lists:

Fig. 6.3 One species from each one of the guilds considered. Top left: stonechat *Saxicola rubicula*, a farmland species; top right: cirl bunting *Emberiza cirlus*, an edge species; bottom left: hawfinch *Coccothraustes coccothraustes*, a forest specialist species; bottom right: blue tit *Cyanistes caeruleus*

the Red Book of Vertebrate of Portugal [13], Species of European Conservation Concern [14] and the International Union for Conservation of Nature (IUCN) Red List for Birds [15].

6.1.3 HNV Features and Explanatory Variables

Due to their nature, *montado* landscapes are distinguished by fuzzy boundaries with overlapping land cover classes (diverse combinations of forest cover, grass and shrubs) [3] (Fig. 6.4). They are often characterized by gradual changes in shrub and tree densities, resulting from the combination of different levels of land use and of variable, extensive land use practices. The fuzziness of the boundaries is inherent to the land use system and should be accepted as such. Small differences in terms of tree density and shrub cover reflect important differences in abiotic factors [16], types of management in the past and present [17], and levels of biodiversity [18].

   We recorded environmental variables for each sampling site that, according to our rationale, could represent the heterogeneity within the *montado*: (1) tree

Fig. 6.4 Different typologies of cork oak *montados* in Portugal: (**a**) medium cover *montado* without grazing; (**b**) dense *montado* without grazing; (**c**) sparse *montado* with cattle; and (**d**) sparse *montado* without grazing

density; (2) shrub density; (3) woody richness; (4) type of edge; and (5) distance to edge (Table 6.2). We also recorded geographical variables (location and altitude), management variables (trunk diameter at breast height, debarking) and variables associated with the sanitary status of the settlements (occurrence of pest outbreaks and fungal presence) (Table 6.2). Immediately after conducting the bird censuses, we evaluated the vegetation features within a 100 m radius around the centre of the point count.

At each point of the study areas, the sanitary status of the oak woodlands was evaluated through visual assessment of five oaks separated 10 m from each other, avoiding trees with adjoined crowns. We detected the damage made by buprestids *Coroebus florentinus* and *Coroebus undatus* through the presence of typical dead branches on the outer canopy and the presence of feeding galleries of larvae on the cork layer, respectively. These measures were used as individual variables, but were also incorporated together with insect trunk holes of bark beetle (Platypodidae and Scolytidae), cerambycids (Cerambycidae) and signs of fungal disease *Biscogniauxia mediterranea* to create a sanitary index (San). We created the sanitary index for *montado* according to the level of harm done by insect pests at the sampling sites. For each insect pest and fungus we attributed a coefficient of impact according to the intensity of their damage on the tree: 1 = high aesthetic

Table 6.2 Environmental variables recorded at each of the sampling sites

Environmental variables	Code
Geographical variables	
Geographical location (coordinates)	Point
Identification of the four areas surveyed	Study Area
Altitude (m)	Alt
Vegetation features	
Shrub density	Shrubs
Tree density	Trees
Woody richness including tree and shrub species (categorical: 1 = 1 or 2 woody species; 2 = 3 or more woody species)	Woody
Management practices	
Trunk diameter at breast height (cm)	DBH
Year of the last cork removal	Debark
Type of edge (categorical: 1 = open area; 2 = shrubs and vineyards; 3 = eucalyptus, pine plantations, orchards, olive groves)	Edge
Distance to the edge (m)	EdgeD
Sanitary status	
Presence/absence of *Biscogniauxia mediterranea*	Bmed
Presence/absence of *Coroebus florentinus* damage	Cflo
Presence/absence of *Coroebus undatus* damage	Cund
Presence/absence of bark beetle (Platypodidae and Scolytidae) and cerambycid (Cerambycidae) damage	BarkB

impact with low economic or ecological relevance (*C. florentinus*); 3 = high economic relevance due to loss of cork value (*C. undatus*); 5 = associated with tree decline or death (bark beetles, cerambycids and *B. mediterranea*).

We created the sanitary index through the sum of the proportion of each pest or disease, which was multiplied by the respective impact coefficient:

$$San = 1\left(\frac{OA\ Cflo}{5\ oaks}\right) + 3\left(\frac{OA\ Cund}{5\ oaks}\right) + 5\left(\frac{OA\ Barkb}{5\ oaks}\right) + 5\left(\frac{OA\ Bmed}{5\ oaks}\right)$$

where OA are oaks affected by each pest or disease.

6.1.4 Data Analysis

Prior to the data analysis and in order to avoid multicollinearity among variables, we performed data reduction procedures. According to Tabachnick and Fidell [19], we assessed all pairwise correlations through Spearman correlation coefficients and retained only one in each pair of highly correlated variables (r > |0.7|) for further analyses. Only a pair of variables showed strong collinearity: altitude and study area ($r = 0.934$). Altitude was removed due to the autocorrelation of this variable between sampling sites belonging to the same study area. We used a one-way analysis of variance (ANOVA) [20] to determine if there were differences between areas in terms of ecological guilds and environmental variables.

We modelled the effects of environmental variables in function of groups of species through linear mixed-effects models [21]. We treated study area as random effect and all other explanatory variables as fixed effects. To deal with model selection uncertainty we analysed our data based on the information theoretic approach (ITA) [22]. In the analyses of species groups (bird richness of every guild), the Shannon diversity index and total species richness, we considered as fixed effects the following six variables, which are representative of the heterogeneity inside the *montado*: shrub density; tree density; woody richness; type of edge; distance to edge; and sanitary index.

We generated all possible models combining from none to seven explanatory variables. We used this option, classified as data dredging [22], because all explanatory variables we considered could potentially influence the response variables. Using all possible combinations, we guaranteed that the explanatory variables were included in the model-averaging procedure in an identical manner. In accordance with Burnham and Anderson [22], we fitted the models one by one and ordered them by their values of AICc (the second-order Akaike information criterion). We used AICc as a measure of information loss for each candidate model, with the best fitting model having the lowest AICc and the highest Akaike weight (wi), which measures the posterior probability of a given model being true, given the data and the set of competing candidate models [22]. Additionally, we also calculated the number of parameters (degrees of freedom), log-likelihood value and AICc difference (ΔAICc), and the model-averaged coefficients of all explanatory variables for each model [22–24]. Finally, we estimated the relative importance of each explanatory variable by adding the Akaike weights of all models in which the variable appeared [22]. The relative importance of the variables that appear in all top models tends towards 1. In variables that only appear in less likely models, their relative importance tends towards 0. We then ranked the explanatory variables according to their relative importance, and the direction and magnitude of the effect of each variable was based on the model-averaged coefficients [22].

We carried out the statistical analysis using SPSS 21 for Windows [25] and R v. 3.0.2 software [26], with the MuMIn package [1] and nlme [2].

6.2 Results

From 74 species recorded in surveys, 50 were used in the analyses, and hereafter we will refer only to those. Almost 40% of species show some level of threat, according to Species of European Conservation Concern (SPEC) [14], or have conservation status according to the Portuguese Red Book of Vertebrates [13] or the IUCN Red List [15, 10]. We detected 17 of the 50 species in all areas. On the other hand, we detected five bird species exclusive to Romeu, three exclusive to Monfurado and Lezírias, and one exclusive to Grândola [10]. The average species richness and standard deviation, per point count, was 6.8 ± 1.3 in Grândola, 7.3 ± 1.2 in Romeu, 8.5 ±2.1 in Lezírias and 11.2 ± 2.4 in Monfurado.

6.2.1 Environmental Variables

In order to evaluate how our environmental variables ranged among the four areas we ran one-way ANOVAs for each variable, with a Bonferroni correction post-hoc test (Table 6.3). The average age of cork removal (Debark) and the distance to the edge (EdgeD) were similar in all areas. Conversely, altitude was significantly different among all areas. Although shrub coverage was higher in Grândola, doubling those in Lezírias and Monfurado, tree coverage and woody diversity were the lowest observed. Monfurado was the area with higher tree coverage and significantly different from all the others. Diversity of woody vegetation was higher in Lezírias and Monfurado and statistically different from Grândola and Romeu. The oldest settlements were observed in Romeu and Lezírias (based on trunk diameter at breast high—DBH), being statistically different from Monfurado.

We observed a similar pattern for both *Coroebus* species, with lower values of affectation in Romeu and higher values in Grândola and Monfurado. In this sense, these areas were significantly different for both pests. Romeu emerged as the area with less evidence of affectation in regard to the sanitary status index.

6.2.2 Bird Guilds

In order to evaluate how our species ranged along the four areas we ran a one-way ANOVA among areas for each species guild, with a Bonferroni correction post-hoc test. Significant differences were observed between areas for all guilds analysed, and also for total bird richness, species diversity and conservation status (Table 6.4). Monfurado was the area with the highest species richness for all bird guilds, with the exception of farmland species. Lezírias had the highest values for this guild, whereas it appears to be residual in Romeu. Romeu had the lowest value for species with conservation status, and Monfurado presented the highest species richness of this guild. As expected, based on the guild results, Monfurado stood out from the other areas regarding species diversity and total species richness. The other areas did not present significant statistical differences, with the exception of higher species richness in Lezírias than in Grândola.

Table 6.3 One-way analysis of variance (ANOVA) for environmental variables

Areas	Shrubs	Trees	Woody	Debark	DBH	Cund	Cflo	San	Alt	EdgeD
G (n = 30)	60 ± 30.73	28 ± 10.64	1.13 ± 0.35	5.13 ± 3.27	35.28 ± 5.76	2.13 ± 0.73	1.70 ± 0.88	7.63 ± 2.35	205.83 ± 43.56	115.27 ± 85.03
L (n = 30)	32.90 ± 27.96	40 ± 15.97	1.87 ± 0.35	6.13 ± 2.54	39.51 ± 8.22	1.73 ± 0.74	1.73 ± 0.91	6.95 ± 1.08	32.27 ± 8.24	86.53 ± 53.26
M (n = 30)	38.50 ± 32.56	55 ± 18.75	1.87 ± 0.35	5.23 ± 1.94	33.16 ± 5.74	2.23 ± 0.86	1.80 ± 0.55	7.31 ± 1.11	294.33 ± 52.89	77.30 ± 76.83
R (n = 27)	52.96 ± 26.86	30.19 ± 11.31	1.15 ± 0.36	6.89 ± 2.59	39.04 ± 7.37	1.48 ± 0.51	1.15 ± 0.36	5.15 ± 1.65	454 ± 91.08	106.41 ± 101.06
F	5.53	20.87	42.08	2.81	5.84	6.67	4.87	12.86	282.58	1.41
P	<0.001	<0.001	<0.001	<0.05	<0.001	<0.001	<0.01	<0.001	<0.001	0.244
Bonferroni correction										
G × R	–	–	–	–	–	<0.01	<0.05	<0.001	<0.001	–
G × L	<0.01	<0.05	<0.001	–	–	–	–	–	<0.001	–
G × M	<0.05	<0.001	<0.001	–	–	–	–	–	<0.001	–
R × L	–	–	<0.001	–	–	–	<0.05	<0.001	<0.001	–
R × M	–	<0.001	<0.001	–	<0.01	<0.01	<0.01	<0.001	<0.001	–
L × M	–	<0.05	–	<0.01	<0.01	–	–	–	<0.001	–

Values represent average richness and respective standard deviation, and results with significant differences after applying the Bonferroni correction

Shrubs density of shrubs, *Trees* density of trees, *Woody* woody richness including trees and shrubs, *Debark* year of the last cork removal, *DBH* trunk diameter at breast high, *Cund* damage by *Coroebus undatus*, *Cflo* damage by *Coroebus florentinus*, *San* sanitary index, *Alt* altitude, *EdgeD* distance to edge according to each area (*G* Serra de Grândola, *L* Companhia das Lezírias, *M* Serra de Monfurado, *R* Romeu)

Table 6.4 One-way analysis of variance (ANOVA) for species richness by ecological guild

Areas	Fa	ES	FoG	FoS	ConSt	ShInd	SpRich
G ($n = 30$)	0.97 ± 1.10	0.50 ± 0.63	4.03 ± 1.16	1.30 ± 0.70	1.20 ± 0.85	0.79 ± 0.10	6.80 ± 1.32
L ($n = 30$)	1.90 ± 1.21	1.20 ± 0.93	4.37 ± 1.38	1.07 ± 0.94	1.57 ± 0.90	0.86 ± 0.12	8.53 ± 2.10
M ($n = 30$)	1.77 ± 1.43	1.80 ± 1.00	5.77 ± 2.37	1.90 ± 1.37	2.23 ± 1.41	0.99 ± 0.15	11.23 ± 2.43
R ($n = 27$)	0.15 ± 0.36	0.85 ± 0.95	5.19 ± 1.39	1.15 ± 0.86	0.63 ± 0.74	0.83 ± 0.08	7.33 ± 1.24
F	14.96	11.69	6.77	4.17	12.62	16.56	33.61
P	<0.001	<0.001	<0.001	<0.01	<0.001	<0.001	<0.001
Bonferroni correction							
G × R	<0.05	–	–	–	–	–	–
G × L	<0.01	<0.05	–	–	–	–	<0.01
G × M	<0.05	<0.001	<0.01	–	<0.01	<0.001	<0.001
R × L	<0.001	–	–	–	<0.01	–	–
R × M	<0.001	<0.01	–	<0.05	<0.001	<0.001	<0.001
L × M	–	–	<0.01	<0.05	–	<0.001	<0.001

Values represent average richness and respective standard deviation, and results with significant differences after applying the Bonferroni correction
Fa farmland species, *ES* edge species, *FoG* forest generalists, *FoS* forest specialists, *ConSt* conservation status, *ShInd* Shannon diversity index, *SpRich* total species richness according to each area (*G* Serra de Grândola, *L* Companhia das Lezírias, *M* Serra de Monfurado, *R* Romeu)

6.2.3   Modelling of Bird Guilds

We ranked the candidate models for each of the response variables based on the AICc (ΔAICc < 2.00) and also estimated the relative importance of each variable (Table 6.4) [22].

6.3   Discussion

In a first approach we characterized areas and associated bird guilds in order to have a better perception of their similarities and differences, allowing us to distinguish possible patterns arising from regional influence. The analysis showed the amplitude of *montados* evaluated throughout the country (Table 6.3), and even within each area, which is particularly relevant when we try to assess associations between bird guilds and HNV features.

The highest richness for most guilds (except for farmland species) was associated with Monfurado (Table 6.5), an area with high amplitude in variables associated with landscape heterogeneity (trees, woody and shrubs; Table 6.3). These results are in line with what has been documented by other authors [4, 6, 27]. Actually, the heterogeneous pattern of wooded matrix with open areas, scattered woodlands and undisturbed patches of Mediterranean forest and scrublands creates a patchwork of habitats, which is a trait of *montado* landscape and induces the highest richness in breeding bird communities in the Iberian Peninsula [5]. So, areas that encompass higher densities of shrubs and trees and that are more diverse in woody species have greater potential to be considered good habitats for a large number of species, including species of conservation concern (*e.g.* redstart *Phoenicurus phoenicurus* and crested tit *Lophophanes cristatus*).

Species belonging to the farmland guild, including thekla lark (*Galerida theklae*), stonechat (*Saxicola rubicula*) and corn bunting *(Emberiza calandra)*, are tolerant of the presence of trees at their breeding sites. Therefore, they can occur in *montados*, contrarily to what happens to strict farmland species [28]. This guild is associated with scattered *montados* and agricultural edges, decreasing its densities with the proximity of forested edges, like eucalyptus or pine plantations. Such a tendency was also recorded by Reino et al. [29] with the same group of species but in a farmland context. According to such relations between birds and habitat, the farmland species occurring in *montados* can be defined as generalist farmland species. Therefore, the diversity of woody vegetation may provide them with more ecological niches; however, shrub coverage must be low since most of these species nest in open ground.

The fuzziness of the *montados* may be the reason for the absence of strong relations between the variables considered and edge species. In *montados* these species are associated with decreasing density of vegetation (*e.g.* trees and shrubs) more than with the abrupt transition between farmland and forestry patches or

Table 6.5 Results from the multimodel inference procedure for the parameters describing species guild associations with environmental variables: relative variable importance (RVI) and trend of the relationship

	Farmland species		Edge species		Forest generalists		Forest specialists		Conservation status		Shannon diversity index		Species richness	
	RVI	Trend	RVI	Trend	RVI	Trend	RVI	Trend	RVI	Trend	RVI	Trend	RVI	Trend
Edge	0.69*	−	0.23	−	0.10	+	0.37*	+	0.89*	−	<0.01	+	0.23	−
Distance to edge	<0.01	+	<0.01	−	<0.01	+	<0.01	+	<0.01	+	<0.01	+	<0.01	−
Sanitary index	0.04	+	0.05	+	0.48*	+	0.05	+	0.06	+	0.99*	+	0.60*	+
Shrub density	0.11		0.01	−	0.01	+	0.33*	+	0.01	−	<0.01	+	0.01	+
Tree density	0.92*	−	0.08	−	0.27*	+	1*	+	0.86*	−	<0.01	+	0.02	+
Woody	0.29*	+	0.17	+	0.42*	+	0.30*	+	0.21	+	0.94*	+	0.96*	+

*Variables included in the best models (second-order Akaike information criterion difference ($\Delta AICc$) < 2.00)

Farmland species: three variables—tree density, edge type and woody species—were included in the set of best models [10]. Tree density assumed the greatest importance (0.92), being negatively associated with the richness of farmland species. Edge type also had high relative importance (0.69), with these species being negatively influenced by the presence of non-agricultural edges. Woody species still had some relative importance. Patches with more woody species showed greater richness of farmland birds

Edge species: none of the variables showed an association with this group of species, the top ranked model being the null model

Forest generalists: although the null model was ranked in the best set of models (in fourth position, (10)), we found it useful to discuss the variables that entered into the other models. These included three variables—sanitary index, woody species and tree density (Table 6.5)—all positively associated with this guild

Forest specialists: the most relevant variables in the candidate models were tree density, edge type, shrub density and woody species. Sites with higher tree cover were most important to this species (1.00), being present in all models

Conservation status: two variables—edge type and tree density—were included in the models, both influencing negatively the richness of species with conservation status. The types of edge that negatively influenced species with conservation status were forest edges (e.g. eucalyptus, pine, orchards) and shrub edges

Shannon diversity index: two variables assumed higher importance and positively influenced species diversity: sanitary index (0.95) and woody species (0.88)

Species richness: as in the case of the Shannon diversity index, the variables included in the models were woody species and sanitary index

early-successional habitats, as in the case of Central and Northern Europe [30, 31]. Additionally, inside the guild there are species associated with different kinds of forest interface. For example, turtle dove (*Streptopelia turtur*), melodious warbler (*Hippolais polyglotta*) and cirl bunting (*Emberiza cirlus*) are mainly associated with forest–farmland edges; wood lark (*Lullula arborea*), Iberian grey shrike (*Lanius meridionalis*) and rock sparrow (*Petronia petronia*) are mainly associated with scattered arboreal vegetation such as sparse *montados*, which can be considered transitional habitats between forested and open areas. The diversity of species–habitat associations in this guild is reflected in the uncertainty of the variables considered in the models [10].

In a more comprehensive view, the forest is the most important component within the cork oak *montados*, and the occurrence of generalist forest species along a wide range of *montado* typologies is to be expected. Our data reveal a preference for diversity of woody vegetation and tree density, although the association with sites with higher pest affectation (reflecting more degraded areas) may indicate the avoidance of sites with higher tree coverage.

The occurrence of forest specialist species was associated with higher densities of trees and shrubs, and with forested edges. These features characterize old and complex *montados*. In a Mediterranean context, species like wren (*Troglodytes troglodytes*), blackcap (*Sylvia atricapilla*) and European robin (*Erithacus rubecula*) use the remaining patches of ancient oak forest with several vegetation strata [32] which is a rare feature across *montados*. Besides, these patches allow an increase in niche availability for species and can be used as an important evaluation element of HNV in Europe [33].

The occurrence of priority species for conservation is one of the key points for a site to be recognized as HNV. In our study these species were associated with sparser *montados* and with farmland edges; in other words, they tended to avoid more mature and complex sites. This trend was directly influenced by the existing unevenness between the species of this group: 16 out of 19 species (84%) with some criteria of conservation were farmland or edge species associated with the forest–farmland interface. We suggest that this result should be considered with some caution because most of the threatened forest species nesting in *montados* were not targeted by our study, since they have large territories (e.g. raptors, black stork), and this could be another factor of bias. The inclusion of other conservation criteria must be considered, such as those proposed by Tavares [34] based on species of which Portugal hosts significant populations at the European level, thus having a responsibility for their conservation (*e.g.* Iberian chiffchaff *Phylloscopus ibericus* or serin *Serinus serinus*).

Species richness and species diversity showed the same trend, being positively associated with areas that have higher diversity of woody vegetation and higher values of sanitary index. The association between these two indices and the sanitary condition may reflect (1) the current state of conservation of *montados* and (2) the influence of the forest generalist species in the species taken as a whole. Most of these species were recorded along several sampling sites (e.g. blue tit *Cyanistes caeruleus*, short-toed treecreeper *Certhia brachydactyla*, blackbird *Turdus merula*), which suggests that the global indices were primarily influenced by the presence of forest generalist species. Cork oak decline has been reported in Southwestern Portugal since the 1890s [35]; therefore, in the light of our results, it seems plausible to state that most of the surveyed areas are under some kind of threat.

6.4 Conclusion

The continuous gradient of land cover and fuzzy boundary characteristic of *montados* [36] are well expressed in the variables of density and diversity of vegetation and their association with the bird guilds evaluated. With the exception of the edge species, these variables were important to all the other bird groups (Table 6.5). The guilds under consideration mainly characterized a gradient of forest complexity, from the farmland species to the forest specialists, and perhaps they can be individually used as HNV indicators of a particular typology of *montado*, based on tree coverage: farmland species for scattered areas and, on the opposite side of the range, forest specialists for more mature settlements or small, well-preserved forest patches. We cannot say if all *montado* typologies can be classified as HNV, but we can define characteristic bird guilds for several typologies, and through the ratio between the species observed and the expected pool of species it should be possible to evaluate if a site may be classified as HNV. At a broader scale the universal measures of species diversity and species richness could also be used as HNV indicators.

The next steps in this research should be focused on: (1) the creation and testing of a compound index with farmland, forest generalist and forest specialist species in order to create a reliable indicator of HNV for *montados* applicable to several scales; (2) the assessment of other HNV parameters for *montados* such as stone piles, ponds or fences; and (3) the integration of other species conservation criteria in addition to the traditional red lists.

References

1. Barton K. Multi-model inference. Version 1.10.0. http://cran.r-project.org/web/packages/MuMIn/index.html. 2014.
2. Pinheiro J, Bates D, DebRoy S, Sarkar D, Team RC. nlme: linear and nonlinear mixed effects models. R package version 3 [Internet]. 2014. p. 1–117. Available from: http://cran.r-roject.org/package=nlme
3. Pinto-Correia T, Vos W. Multifunctionality in Mediterranean landscapes—past and future. New Dimens Eur Landsc. 2004;4:135–64.
4. Blondel J, Aronson J. Biology and wildlife of the Mediterranean region. New York: Oxford University Press; 1999.
5. Tellería J. Passerine bird communities of Iberian *dehesas*: a review. Anim Biodivers Conserv. 2001;24(2):67.

6. Díaz M, Pulido F, Marañón T. Diversidad biológica y sostenibilidad ecológica y económica de los sistemas adehesados. Ecosistemas XII. 2003;

7. Telleria J, Baquero R, Santos T. Effects of forest fragmentation on European birds: implications of regional differences in species richness. J Biogeogr. 2003;30:621–8.

8. Pereira P, Godinho C, Roque I, Marques A, Branco M, Rabaça JE. Time to rethink the management intensity in a Mediterranean oak woodland: the response of insectivorous birds and leaf-chewing defoliators as key groups in the forest ecosystem. Ann For Sci. 2014;71 (1):25–32.

9. Pereira P, Godinho C, Roque I, Rabaça JE. O Montado e as Aves: boas práticas para uma gestão sustentável. Câmara Municipal de Coruche e Universidade de Évora. 2015.

10. Catarino L, Godinho C, Pereira P, Luís A, Rabaça JE. Can birds play a role as High Nature Value indicators of *montado* system? Agrofor Syst. 2014;90(1):45–56.

11. Carreiras J, Pereira J, Pereira J. Estimation of tree canopy cover in evergreen oak woodlands using remote sensing. For Ecol Manag. 2006;223:45–53.

12. Bibby C, Burgess N, Hill D, Mustoe S. Bird census techniques. 2nd ed. London: Academic; 2000.

13. Cabral M, Almeida J, Almeida P, Dellinger T, Almeida N, Oliveira M, et al. Livro Vermelho dos Vertebrados de Portugal. Lisboa: Instituto da Conservação da Natureza; 2005.

14. BirdLife International. Birds in the European Union: a status assessment. Wageningen: BirdLife International; 2004.

15. BirdLife International. IUCN red list for birds [Internet]. BirdLife International. 2014. Available from: http://www.birdlife.org

16. Joffre R, Rambal S, Ratte J. The *dehesa* system of southern Spain and Portugal as a natural ecosystem mimic. Agrofor Syst. 1999;45(1–3):57–79.

17. Pinto-Correia T. Threatened landscape in Alentejo, Portugal: the "*montado*" and other "agro-silvo-pastoral" systems. Landsc Urban Plan. 1993;24(1–4):43–8.

18. Ojeda F, Arroyo J, Marañón T. Biodiversity components and conservation of Mediterranean heathlands in Southern Spain. Biol Conserv. 1995;72(1):61–72.

19. Tabachnick B, Fidell L. Using multivariate statistics. Boston: Allyn and Bacon; 2001.

20. Zar JH. Biostatistical analysis. 4th ed. Englewood Cliffs: Prentice Hall; 1999.

21. Pinheiro J, Bates D. Mixed-effects models in S and S-Plus. Springer, editor. New York; 2000.

22. Burnham K, Anderson D. Model selection and multimodel inference: a practical information-theoretic approach. New York: Springer; 2002.

23. Lukacs P, Burnham K, Anderson D. Model selection bias and Freedman's paradox. Ann Inst Stat Math. 2010;62:117–25.

24. Symonds MR. A brief guide to model selection, multimodel inference and model averaging in behavioural ecology using Akaike's information criterion. Behav Ecol Sociobiol. 2011;65 (1):13–21.

25. IBMCorp. IBM SPSS statistics for windows. Version 21.0. IBM Corp., Armonk, NY. 2012.

26. RCoreTeam. A language and environment for statistical computing. R Foundation for Statistical Computing. Vienna, Austria; 2013.

27. Godinho C, Rabaça JE. Birds like it corky: the influence of habitat features and management of "*montados*" in breeding bird communities. Agrofor Syst. 2011;82(2):138–95.

28. Moreira F, Beja P, Morgado R, Reino L, Gordinho L, Delgado A, et al. Effects of field management and landscape context on grassland wintering birds in Southern Portugal. Agric Ecosyst Environ. 2005;109(1–2):59–74.

29. Reino L, Beja P, Osborne P, Morgado R, Fabião A, Rotenberry J. Distance to edges, edge contrast and landscape fragmentation: interactions affecting farmland birds around forest plantations. Biol Conserv. 2009;142(4):824–38.

30. Imbeau L, Drapeau P, Mönkkönen M. Are forest birds categorised as "edge species" strictly associated with edges? Ecography (Cop). 2003;26(4):514–20.

31. Storch I, Woitke E, Krieger S. Landscape-scale edge effect in predation risk in forest–farmland mosaics of Central Europe. Landsc Ecol. 2005;20(8):927–40.

32. Pérez-Tris J, Bensch S, Carbonell R, Helbig A, Tellería J. Historical diversification of migration patterns in a passerine bird. Evolution (N Y). 2004;58(8):1819–32.
33. Morelli F. Relative importance of marginal vegetation (shrubs, hedgerows, isolated trees) surrogate of HNV farmland for bird species distribution in Central Italy. Ecol Eng. Elsevier B. V.; 2013 Aug;57(July):261–6.
34. Tavares JP. Uma perspectiva internacional sobre as prioridades de conservação da avifauna Portuguesa. In: Actas do VI Congresso de Ornitologia da Sociedade Portuguesa para o Estudo das Aves Elvas. 2009.
35. Cabral M, Ferreira M, Moreira T, Carvalho E, Diniz A. Diagnóstico das causas da anormal mortalidade dos sobreiros a Sul do Tejo. Sci Gerund. 1992;18:205–14.
36. Pinto-Correia T, Ribeiro N, Sá-Sousa P. Introducing the *montado*, the cork and holm oak agroforestry system of Southern Portugal. Agrofor Syst. 2011;82(2):99–104.

Chapter 7
Case Study 3. Using Indicator Species Analysis IndVal to Identify Bird Indicators of HNV in Farmlands from Western Poland

Piotr Tryjanowski and Federico Morelli

Abstract

- Country: Poland
- Year of study: 2010
- Dominant environment in study area: croplands
- Ecoregion: Central European mixed forests
- Climate: temperate climate
- Target indicator: set of common bird species
- Main statistical tools: indicator analysis
- Some useful R packages: 'indicspecies' [1]

Summary Case Study No. 3

- Country: Poland
- Year of study: 2010
- Dominant environment in study area: croplands
- Ecoregion: Central European mixed forests
- Climate: temperate climate
- Target indicator: set of common bird species
- Main statistical tools: indicator analysis
- Some useful R packages: 'indicspecies' [1]

Farmland landscapes are recognized as important ecosystems, not only for their rich biodiversity but equally so for the human beings who live and work in these places.

P. Tryjanowski (✉)
Institute of Zoology, Poznań University of Life Sciences, Wojska Polskiegeo 71 C, PL-60-625 Poznań, Poland
e-mail: piotr.tryjanowski@gmail.com

F. Morelli
Faculty of Environmental Sciences, Department of Applied Geoinformatics and Spatial Planning, Czech University of Life Sciences Prague, Kamýcká 129, CZ-165 00 Prague 6, Czech Republic

© Springer International Publishing AG 2017
F. Morelli, P. Tryjanowski (eds.), *Birds as Useful Indicators of High Nature Value Farmlands*, DOI 10.1007/978-3-319-50284-7_7

In this case study, performed in 2010 in Western Poland [2, 3], we examined the relationships between High Nature Value (HNV) farmlands and the occurrence of bird species, by means of indicator analysis, applying the IndVal tool in order to identify bird species 'indicators' of HNV farmland and bird species indicators of non-HNV farmlands. Furthermore, we examined the effects of HNV farmlands on the taxonomic diversity of bird communities (bird species richness).

7.1 Methodology

7.1.1 Study Area

The study was conducted in an agricultural landscape of Western Poland, near Odolanów (51° 34′ N, 17° 40′ E) (Fig. 7.1). The study location (38,000 ha) is an extensively used agricultural landscape and comprises a mosaic of meadows and pastures (44%), arable fields (42%) and midfield woodlots of different ages (6%), plus scattered trees and discontinuous linear habitats, mainly consisting of mixed rows of trees and shrubs (see details in [4]).

The climate in farmlands of Western Poland is a temperate climate. This area was selected because it contains a large variety of different farming practices, with good examples of HNV areas (see Fig. 7.2, photos).

7.1.2 Species and Environmental Data

A total of 51 sample sites in agricultural landscapes were visited at least two times during the breeding season in the year 2010. Sampling involved taking 5 min point counts at each sample site, extending from half an hour after sunrise until 4.5 h after sunrise. Counts were taken only during favourable weather conditions. Point counts provide highly reliable estimates of relative population density, and this is a standardized and practical method for comparing bird communities between different habitats and times [5, 6].

At each sample site, the species richness was calculated as the maximum number of birds recorded, considering the two visits.

All sites were classified in situ, using a binary system to categorize each farmland as HNV [1] or non-HNV (0), following the classification proposed in Andersen et al. [7].

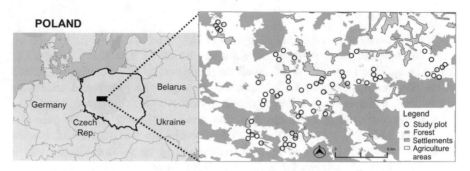

Fig. 7.1 Study area and sampled sites in Western Poland

Fig. 7.2 Farming types in Western Poland: (**a–d**) some examples of High Nature Value farmland characterized by the presence of isolated trees, forest margins and scattered shrubs (Photo (**a**): A. Wuczyński)

7.1.3   Data Analysis

1. Preliminary Comparisons

Difference in bird species richness in different farmland categories (HNV and non-HNV) was assessed graphically using boxplots and statistically using a Wilcoxon test because the data followed a non-normal distribution [8].

2. Species Indicator Analysis

In order to study the bird species typical of HNV and non-HNV farmlands, we used the indicator value method (IndVal analysis) [1, 9]. The IndVal analysis is based on specificity, which is the conditional probability of a positive predictive value of a given species as an indicator of the target plot group, and sensitivity (or fidelity), which is the conditional probability that the given species will be found in a newly surveyed plot belonging to the same plot group [10], producing a percentage indicator value (IndVal) for each species. In this study, bird species having an IndVal statistic value higher than 20% and a p value lower than 0.05 were considered indicator species for each category of farmland [11]. IndVal analysis was performed using the 'indicspecies' package [1].

All tests were performed with the R program [12].

7.2 Results

7.2.1 Farmland Classification and Description

Of the 51 farmland sites surveyed, 43 of them (59.7%) were located in HNV areas. Bird species richness was slightly higher in HNV farmland (mean of bird richness 17.4, range 6–32) than in non-HNV farmland (15.0, range 5–30) (t = 1.20, df 45.3, $p < 0.05$). Also, the maximum values of species richness was recorded in HNV farmland (32 species).

7.2.2 Bird Indicators of HNV Farmland

Four bird species were found as indicators of HNV farmlands (Fig. 7.3), and two bird species were significantly diagnostic of non-HNV farmlands in Western Poland. The complete output, with the results of the IndVal analysis, is provided in Table 7.1.

7.3 Discussion

The bird species positively related to HNV presence—*Anthus pratensis, Gallinago gallinago, Emberiza schoeniclus* and *Crex crex*—are known as farmland specialists, related mainly to semi-natural meadows and pastures [13–16]. On the other hand, the bird species positively related to non-HNV areas were *Phoenicurus ochruros, Erithacus rubecula* and, less significantly, *Passer domesticus and Riparia riparia*. The first three species are related to village infrastructure,

Fig. 7.3 Four bird species correlated with High Nature Value (HNV) farmlands in Western Poland. (**a**) *Anthus pratensis*; (**b**) *Gallinago gallinago*; (**c**) *Emberiza schoeniclus*; (**d**) *Crex crex* (Photo (**c**): C. Korkosz)

Table 7.1 IndVal analysis results on bird species indicators of High Nature Value (HNV) and non-HNV farmlands in Western Poland

Group HNV farmlands, number of species = 4				
Bird species	**A**	**B**	**stat**	**p value**
Anthus pratensis	0.923	0.461	0.653	0.005**
Gallinago gallinago	0.822	0.500	0.641	0.010**
Emberiza schoeniclus	0.787	0.308	0.492	0.050*
Crex crex	0.980	0.192	0.439	0.050*
Group non-HNV farmlands, number of species = 4				
Bird species	**A**	**B**	**stat**	**p value**
Phoenicurus ochruros	0.844	0.333	0.530	0.035*
Erithacus rubecula	0.812	0.292	0.487	0.045*
Passer domesticus	0.791	0.292	0.480	0.070[†]
Riparia riparia	0.999	0.125	0.354	0.080[†]

Component A refers to the specificity, while component B refers to the fidelity of the species to each farmland category. Only species with stat values higher than 0.20 were selected at a p value of 0.05
Significance codes: [†] $p \le 0.1$; * $p \le 0.05$; ** $p \le 0.01$

including small green patches [17]; the final one prefers mainly sandy dunes, sometimes also very close to human settlements, where it has a possibility to prepare a breeding colony in sandy banks [18]. What is more important, the birds from the first group belong to declined open habitat–related species at the whole-Europe continental scale [19, 20]. Therefore, we have to pay attention to the obtained results with caution, and not just say that half of the study species are related to HNV and half are not. It seems that the crucial factor is the water regime in the study area, which especially is a factor in relatively dry farmland habitats and has already been described as crucial factor also for the studied species [16, 21]. Importantly, in Western Poland a very serious problem has occurred with water levels due to bad irrigation projects, climate change and irrational management during seasons [22].

Generally, negative factors that play an important role at the European scale, and that are also important in the study area, include land use change from meadows and pastures to arable fields, increased use of pesticides and fertilizers, and habitat fragmentation [23]. However, local farmlands in Poland support extremely rich bird communities, which contrasts markedly to farmland habitats in Western Europe [18, 23]. There are probably different reasons for this high bird species richness. First, the study area is a place with extensive agriculture and is located in a complex landscape in a river valley, which attracts large numbers of the common species [18]. It is underlined by slightly higher bird species diversity in HNV places than in non-HNV areas. These differences can also be explained by the naturalness of the habitats in relation to the water regime, but the situation is probably more complex. One of the factors connected with HNV farmland is also pastoralism and general openness of the habitat. Both factors affect food availability, as well as being associated with a lower predation rate than more heterogeneous habitats with a matrix of microhabitats [24–26].

To conclude, a slight but positive effect of HNV areas on the overall species richness was found in farmland in Poland. This is an important aspect that can encourage planning based on the use of a network of protected areas as a conservation tool [27, 28]. However, the effect differs between species. We suggest that using the HNV farmland concept can constitute a useful tool for effective conservation planning.

References

1. De Cáceres M, Jansen F. "indicspecies" R package-functions to assess the strength and significance of relationship of species site group associations; 2016.
2. Kwieciński Z, Morelli F, Antczak M, Hromada M, Szymański P, Tobolka M, Jankowiak Ł, Tryjanowski P. Seasonal changes in avian communities living in an extensively used farmland of Western Poland. Eur J Ecol. (2017);2:105–16
3. Tryjanowski P, Morelli F. Presence of cuckoo reliably indicates high bird diversity: a case study in a farmland area. Ecol Indic [Internet]. Elsevier Ltd; 2015;55:52–8. Available from: http://linkinghub.elsevier.com/retrieve/pii/S1470160X15001363

4. Hromada M, Tryjanowski P, Antczak M. Presence of the great grey shrike *Lanius excubitor* affects breeding passerine assemblage. Ann Zool Fenn. 2002;39:125–30.
5. Bibby CJ, Burgess ND, Hill DA. Bird census techniques (Google eBook) [Internet]. Academic Press;1992 [cited 2014 Apr 24]. 257 p. Available from: http://books.google.com/books?id=5TqfwEHCVuoC&pgis=1
6. Voříšek P, Klvaňová A, Wotton S, Gregory RD. A best practice guide for wild bird monitoring schemes. Brussels: Pan-European Common Bird Monitoring Scheme (PECMBS); 2010. 151 p.
7. Andersen E, Baldock D, Bennett H, Beaufoy G, Bignal EM, Brouwer F, et al. Developing a high nature value indicator. Report for the European Environment Agency [Internet]. Copenhagen; 2003. Available from: http://www.ieep.eu/assets/646/Developing_HNV_indicator.pdf
8. Bauer DF. Constructing confidence sets using rank statistics. J Am Stat Assoc. 1972;67:687–90.
9. De Cáceres M, Legendre P, Moretti M. Improving indicator species analysis by combining groups of sites. Oikos [Internet]. 2010e Oct 14 [cited 2014 Jan 21];119(10):1674–84. Available from: http://doi.wiley.com/10.1111/j.1600-0706.2010.18334.x
10. Dufrene M, Legendre P. Species assemblages and indicator species: the need for a flexible asymmetrical approach. Ecol Monogr. 1997;67:345–66.
11. Della Rocca F, Stefanelli S, Pasquaretta C, Campanaro A, Bogliani G. Effect of deadwood management on saproxylic beetle richness in the floodplain forests of northern Italy: some measures for deadwood sustainable use. J Insect Conserv [Internet]. 2014 Feb 13 [cited 2014 Dec 13];18(1):121–36. Available from: http://link.springer.com/10.1007/s10841-014-9620-1
12. R Development Core Team. R: a language and environment for statistical computing [Internet]. Vienna, Austria: R Foundation for Statistical Computing, Vienna, Austria; 2016. Available from: http://www.r-project.org
13. Kuczynski L, Osiejuk TS, Tryjanowski P. Bird–habitat relationships on wet meadows in the Slonsk Nature Reserve W Poland. Biol Bull Poznań. 2000;37(2):257–66.
14. Herzon I, Auninš A, Elts J, Preikša Ž. Habitat associations of farmland birds across the East Baltic region. Acta Zool Litu. 2006;16(4):249–60.
15. Budka M, Osiejuk TS. Habitat preferences of corncrake (*Crex crex*) males in agricultural meadows. Agric Ecosyst Environ. 2013;171:33–8.
16. Kosicki JZ, Chylarecki P. The hooded crow *Corvus cornix* density as a predictor of wetland bird species richness on a large geographical scale in Poland. Ecol Indic [Internet]. 2014 Mar [cited 2014 May 4];38:50–60. Available from: http://www.sciencedirect.com/science/article/pii/S1470160X13004032
17. Rosin ZM, Skórka P, Pärt T, Żmihorski M, Ekner-Grzyb A, Kwieciński Z, et al. Villages and their old farmsteads are hot spots of bird diversity in agricultural landscapes. J Appl Ecol. 2016;53(5):1363–72.
18. Skórka P, Mrtyka R, Wójcik JD. Species richness of breeding birds at a landscape scale—which habitat is the most important? Acta Ornithol. 2006;41(1):49–54.
19. Sanderson FJ, Kloch A, Sachanowicz K, Donald PF. Predicting the effects of agricultural change on farmland bird populations in Poland. Agric Ecosyst Environ. 2009;129(1):37–42.
20. Ławicki Ł, Wylegała P, Batycki A, Kajzer Z, Guentzel S, Jasiński M, et al. Long-term decline of grassland waders in western Poland. Vogelwelt. 2011;132:101–8.
21. Brickle NW, Peach WJ. The breeding ecology of reed buntings *Emberiza schoeniclus* in farmland and wetland habitats in lowland England. Ibis (Lond 1859). 2004;146(s2):69–77.
22. Juszczak R, Kędziora A, Olejnik J. Assessment of water retention capacity of small ponds in Wyskoć agricultural-forest catchment in Western Poland. Pol J Environ Stud. 2007;16(5):685–95.
23. Tryjanowski P, Hartel T, Báldi A, Szymański P, Tobółka M, Herzon I, et al. Conservation of farmland birds faces different challenges in Western and Central–Eastern Europe. Acta Ornithol [Internet]. 2011 Jun [cited 2013 Feb 19];46(1):1–12. Available from: http://www.bioone.org/doi/abs/10.3161/000164511X589857

24. Evans KL. The potential for interactions between predation and habitat change to cause population declines of farmland birds. Ibis (Lond 1859). 2004;146(1):1–13.
25. Paillisson JM, Reeber S, Marion L. Bird assemblages as bio-indicators of water regime management and hunting disturbance in natural wet grasslands. Biol Conserv. 2002;106(1): 115–27.
26. Vukelič E. Effects of meadow management practices on the breeding birds of Ljubljansko Barje (Central Slovenia). Acrocephalus. 2009;30(140):3–15.
27. Grodzińska-Jurczak M, Cent J. Expansion of nature conservation areas: problems with Natura 2000 implementation in Poland? Environ Manage [Internet]. 2011 Jan [ecited 2014 Oct 26];47(1):11–27. Available from: http://www.pubmedcentral.nih.gov/articlerender.fcgi? artid=3016195&tool=pmcentrez&rendertype=abstract
28. Wuczyński A. Farmland bird diversity in contrasting agricultural landscapes of southwestern Poland. Landsc Urban Plan. 2016;148:108–19.

Chapter 8
Discussion and Final Considerations

Federico Morelli and Piotr Tryjanowski

Abstract In this chapter, we discuss briefly some issues and limitations related to model bird species and High Nature Value (HNV) farmlands. We mention spatial autocorrelation (SAC) issues and necessary corrections in the modelling procedure, and the alternative of the multimodel inference (MMI) approach when working on models with many covariates. We discuss the importance of the transferability or adaptability of models. Finally, we highlight the potentialities related to the use of few common species as indicators for monitoring programmes: use of a citizen science approach as a cost-effective monitoring tool.

Keywords Model selection criteria • Model limitations • Transferability-adaptability • SAC • Citizen Science

8.1 Model Issues and Limitations: Transferability or Adaptability of Models?

Even if models in combination with the bioindicator approach constitute a particularly useful tool for monitoring programmes and ecological planning, as well as for studying and classifying agro-ecosystems, it is necessary to mention some potential concerns related to the proposed methodology.

F. Morelli (✉)
Faculty of Environmental Sciences, Department of Applied Geoinformatics and Spatial Planning, Czech University of Life Sciences Prague, Kamýcká 129, CZ-165 00, Prague 6, Czech Republic
e-mail: fmorellius@gmail.com

P. Tryjanowski
Institute of Zoology, Poznań University of Life Sciences, Wojska Polskiego 71 C, PL-60-625, Poznań, Poland

© Springer International Publishing AG 2017   115
F. Morelli, P. Tryjanowski (eds.), *Birds as Useful Indicators of High Nature Value Farmlands*, DOI 10.1007/978-3-319-50284-7_8

8.1.1   Spatial Autocorrelation Issues

First of all, as for every modelling procedure focused on ecological data, it is necessary to consider the potential bias in model outputs related to the presence of spatial autocorrelation (SAC) in the data [1, 2]. SAC is common in ecological or biological data (as used in the methodologies proposed in this book). In a few words, SAC refers to spatial dependency or covariation of data properties in relationship to the geographical space among these data.

We know that point patterns in space can be studied in order to recognize if events are systematically organized or clumped compared with events distributed at random [3], and we can use this knowledge in order to detect SAC in our data set. A good strategy is to use Mantel or Moran'I tests [4–6] to compare the spatial congruence or mismatch between two matrices derived from our data set. The Mantel test can evaluate the similarity between these two matrices, one measuring ecological distance (the difference in the response variable values among sites) and the other measuring the geographical distance among the same sites [7]. When spatial autocorrelation exists, then the closer the plots are in geometric space, the more similar the pattern of values between matrices should be [8]. It is possible to use Monte Carlo permutations with randomizations to test for Mantel test statistical significance, applying the 'vegan' package for R [9].

If SAC is present in the data set, many procedures are described to remove or limit this issue in the performance of models. For example, in order to avoid any possible SAC issues, geographical coordinates of sample sites can be added as covariates, or one could incorporate a spatial term into the modelling procedure. However, for a complete and detailed explanation about the methods for taking into account SAC problems, we suggest to read the following papers [10, 11].

8.1.2   Overfitting Risks in Models with Many Covariates

One problem when modelling many bird species as fixed factors, trying to find species indicators of HNV farmlands, is related to same problem detected for models with too many covariates: they can be affected by overfitting risks [12]. In this regard, we suggest working with a large data set, with more than ten observations for each parameter (fixed factor, or bird species in this particular case) considered in the full model. As an alternative, we suggest applying a multimodel inference (MMI) approach or other alternatives when modelling a large number of covariates [13].

8.1.3 Selection of Bird Indicators in Different HNV Types

With regard to the selection of potential bird species as candidates to be indicators of HNV farming systems, we point out how it is important to consider the following aspects:

(a) Separately applying the species distribution model (SDM) procedure and bioindicator for each type of HNV, because bird communities inhabiting farmland or grasslands are different [14], and then each kind of HNV farming systems have to be addressed using different set of bird indicators

(b) Selecting species with an occurrence equal to or higher than a given threshold defined ad hoc for each particular case study, avoiding rare species or too ubiquitous species, because they are not adequate as bioindicators [15, 16]

(c) Preferring a multispecies approach rather than a single-species approach, because even a few species can define more roles in bird communities and then be suitable to better characterize complex environments, as HNV farming systems are

(d) Considering the habitat or diet specialization of bird species, which can play an important role [17]

8.1.4 Spatial and Time Scales and Adaptability of Models

For these kinds of models, model robustness tends to be more accurate at a local or regional scale [18], and so it is strongly suggested to develop different models for each different region. Form this point of view, we consider that 'adaptability' is more important than 'transferability' of models regarding the proposed methodology.

Furthermore, models need to be calibrated periodically, because the predictive power of even the best models can change over time, due to fluctuations related to the ecology of indicator species that are not necessary correlated with the characteristics of HNV farming systems.

Finally, it is important to be cautious in ecological comparisons between farmlands considered to be HNV and non-HNV when focusing on the biodiversity of bird communities supported for each one, since a larger number of species observed in HNV farms may be due to an increase in the number of habitat-generalist species, thus decreasing the functional diversity of the overall bird community. Thus, for any comparison of the biodiversity of HNV and non-HNV farming systems, we suggest considering not only taxonomic diversity measures (bird species richness), but also the composition of the bird community, in terms of farmland specialists and habitat generalists, then considering bird traits, to be able to better describe the effects of farming systems on each component of biodiversity, as explained in Chap. 4 of this book.

8.2 Potentialities Related to the Use of a Few Common Species as Indicators for Monitoring Programmes: A Citizen Science Approach

Our findings support the use of bioindicator species in monitoring programmes for agro-ecosystems in Europe. And considering the main criteria for the definition of a good surrogate (a balance between robustness, communicability, accuracy, generality, cost effectiveness and good transferability of the surrogacy) [19], we highlight the high potentialities shown for the proposed framework.

Furthermore, because much of the work in biodiversity conservation is carried out by non-governmental conservation organizations with public support, the use of a few common species is expected to be a valuable tool for communication and for convincing the public, authorities and politicians to preserve and enhance biodiversity in these particular landscapes [20].

The identification of a few common bird species—species that are easy to recognize and then monitor—can offer an unique possibility to set up citizen programmes focused on ecological planning [21].

The proposed methodology is a good candidate for a biological survey of agro-ecosystems enlisting ordinary citizens to monitor the status of HNV farmlands along a temporal dimension. Some interesting examples of the use of citizen programmes have come from France, where one operation was conducted, for instance, in 2012. In this particular initiative, a short TV advertisement invited young people to detect cuckoo arrivals and participate in a web survey. An example can be seen at http://www.dailymotion.com/playlist/x1yf6c_yannaki_missions-printemps-2012/1#video=xpon1m

The use of volunteers for data collection is an efficient strategy to collect high-quality data, with large savings for public institutions, as demonstrated in many studies [22–26].

Therefore, citizen science provides a great modality to facilitate people's engagement and learning in science, as well as an opportunity for direct participation in topics of wide interest for the whole population.

References

1. Sokal RR, Oden NL. Spatial autocorrelation in biology: 2. Some biological implications and four applications of evolutionary and ecological interest. Biol J Linn Soc. 1978;10:229–49.
2. Valcu M, Kempenaers B. Spatial autocorrelation: an overlooked concept in behavioral ecology. Behav Ecol. 2010;21(5):902–5.
3. Moore DA, Carpenter TE. Spatial analytical methods and geographic information systems: use in health research and epidemiology. Epidemiol Rev [Internet]. 1999 Jan 1;21(2):143–61. Available from: http://epirev.oxfordjournals.org/cgi/doi/10.1093/oxfordjournals.epirev.a017993
4. Legendre P, Gauthier O. Statistical methods for temporal and space–time analysis of community composition data statistical methods for temporal and space–time analysis of community composition data. Proc R Soc London B Biol Sci [Internet]. 2014;281:20132728. Available from: http://dx.doi.org/10.1098/rspb.2013.2728

5. Mantel N. The detection of disease clustering and a generalized regression approach. Cancer Res. 1967;27:209–20.

6. Diniz-Filho JAF, Soares TN, Lima JS, Dobrovolski R, Landeiro VL, de Campos Telles MP, et al. Mantel test in population genetics. Genet Mol Biol [Internet]. 2013 Dec [cited 2016 May 4];36(4):475–85. Available from: http://www.pubmedcentral.nih.gov/articlerender.fcgi?artid=3873175&tool=pmcentrez&rendertype=abstract

7. Guillot G. On the use of the simple and partial Mantel tests in presence of spatial autocorrelation. Math Model [Internet]. 2011;2011:1–13. Available from: http://arxiv.org/abs/1112.0651

8. Legendre P, Legendre L. Numerical ecology [Internet]. 3rd ed. Amsterdam: Elsevier; 2012 [cited 2014 Jul 23]. 1006 p. Available from: https://www.elsevier.com/books/numerical-ecology/legendre/978-0-444-53868-0

9. Oksanen J, Guillaume Blanchet F, Kindt R, Legendre P, Minchin PR, O'Hara BR, et al. vegan: community ecology package. R package version 2.3-4 [Internet]. 2016. p. 291. Available from: https://cran.r-project.org/package=vegan

10. Dormann CF, McPherson JM, Araújo MB, Bivand R, Bolliger J, Carl G, et al. Methods to account for spatial autocorrelation in the analysis of species distributional data: a review. Ecography (Cop) [Internet]. 2007 Oct 27 [cited 2014 Jul 10];30(5):609–28. Available from: http://doi.wiley.com/10.1111/j.2007.0906-7590.05171.x

11. Legendre P. Spatial autocorrelation: trouble or new paradigm? Ecology. 1993;74(6):1659–73.

12. Stokes ME, Davis CS, Koch GG. In: Cary N, editor. Categorical data analysis using the SAS System. 2nd ed. Cray: SAS Publishing/BBU Press/Wiley; 2009 .648 p

13. Guillera-Arroita G, Lahoz-Monfort JJ, Elith J, Gordon A, Kujala H, Lentini P, et al. Is my species distribution model fit for purpose? Matching data and models to applications. Glob Ecol Biogeogr. 2015;24(3):276–92.

14. Aue B, Diekötter T, Gottschalk TK, Wolters V, Hotes S. How High Nature Value (HNV) farmland is related to bird diversity in agro-ecosystems—towards a versatile tool for biodiversity monitoring and conservation planning. Agric Ecosyst Environ [Internet]. Elsevier B.V.; 2014;194:58–64. Available from: http://dx.doi.org/10.1016/j.agee.2014.04.012

15. Padoa-Schioppa E, Baietto M, Massa R, Bottoni L. Bird communities as bioindicators: the focal species concept in agricultural landscapes. Ecol Indic [Internet]. 2006 Jan [cited 2013 Jun 5];6(1):83–93. Available from: http://linkinghub.elsevier.com/retrieve/pii/S1470160X05000671

16. Holt EA, Miller SW. Bioindicators: using organisms to measure environmental impacts. Nat Educ Knowl. 2010;3(10):8.

17. Kujawa K. Population density and species composition changes for breeding bird species in farmland woodlots in Western Poland between 1964 and 1994. Agric Ecosyst Environ. 2002;91:261–71.

18. Morelli F, Jerzak L, Tryjanowski P. Birds as useful indicators of High Nature Value (HNV) farmland in Central Italy. Ecol Indic. 2014;38:236–42.

19. Lindenmayer DB, Pierson J, Barton PS, Beger M, Branquinho C, Calhoun A, et al. A new framework for selecting environmental surrogates. Sci Total Environ [Internet]. Elsevier B.V.; 2015;538:1029–1038. Available from: http://linkinghub.elsevier.com/retrieve/pii/S0048969715305593

20. Home R, Keller C, Nagel P, Bauer N, Hunziker M. Selection criteria for flagship species by conservation organizations. Environ Conserv [Internet]. 2009 Aug 14 [cited 2014 Oct 30];36(2):139. Available from: http://www.journals.cambridge.org/abstract_S0376892909990051

21. Kobori H, Dickinson JL, Washitani I, Sakurai R, Amano T, Komatsu N, et al. Citizen science: a new approach to advance ecology, education, and conservation. Ecol Res [Internet]. Springer Japan; 2016;31(1):1–19. Available from: http://dx.doi.org/10.1007/s11284-015-1314-y

22. Jiguet F, Devictor V, Julliard R, Couvet D. French citizens monitoring ordinary birds provide tools for conservation and ecological sciences. Acta Oecologica [Internet]. Elsevier Masson SAS; 2012 Oct [cited 2014 Oct 15];44:58–66. Available from: http://linkinghub.elsevier.com/retrieve/pii/S1146609X11000762

23. Cooper CB, Dickinson J, Phillips T, Bonney R. Citizen science as a tool for conservation in residential ecosystems. Ecol Soc [Internet]. 2007;12(2):11. Available from: http://www. ecologyandsociety.org/vol12/iss2/art11/

24. Roy HE, Baxter E, Saunders A, Pocock MJO. Focal plant observations as a standardised method for pollinator monitoring: opportunities and limitations for mass participation citizen science. Ollerton J, editor. PLoS One [Internet]. 2016 Mar 17 [cited 2016 Jul 9];11(3): e0150794. Available from: http://dx.plos.org/10.1371/journal.pone.0150794

25. Devictor V, Whittaker RJ, Beltrame C. Beyond scarcity: citizen science programmes as useful tools for conservation biogeography. Divers Distrib [Internet]. 2010 Apr 13 [cited 2014 Jul 9];16(3):354–62. Available from: http://doi.wiley.com/10.1111/j.1472-4642.2009.00615.x

26. McCaffrey RE. Using citizen science in urban bird studies. Urban Habitats [Internet]. 2005;3(1):70–86. Available from: http://www.urbanhabitats.org

Printed in the United States
By Bookmasters